Machine Shop Trade Secrets

A GUIDE TO MANUFACTURING MACHINE SHOP PRACTICES

James A. Harvey

Industrial Press Inc.

989 Avenue of the Americas, New York, NY 10018

Library of Congress Control Number: 2003095415

ISBN 0-8311-3227-2

Comments, criticisms and suggestions are invited, and should be forwarded to ProShop Publishing, Attn: James A. Harvey, 12112 St. Mark St., Garden Grove, CA 92845

First Industrial Press Edition, March 2005

Manufactured in the U.S.A.

Design: White Light Publishing

Visit us at our website:

www.industrialpress.com

10 9 8 7 6

Contents

List of Figures

Acknowledgements

I'd like to thank my wife Reyna for putting up with the long hours I've spent in front of the computer instead of with her, my daughter Joanna who inspired me to stick with the project and my son Billy for the hours of entertainment he's provided.

I am grateful for the support of my cousin Ann Fahl who did much of the editing, Betsy Grey for doing the illustrations, and my parents who have given me, among many great things, the camera used to shoot the photos in this book.

And a salute to some of my direct and indirect mentors, Bob Pequinot, Bill Glaze, Mike Davis (deceased), Eugene Sterncorb, Bill Pfister, Juvencio Arancibia, Paul Hudson, Steve Bui, and Gregg Young.

Introduction

There are many reference books on the market today, starting with *Machinery's Handbook,* that provide much of the technical and reference information a machinist or engineer may ever need. It is difficult, however, to find a book that provides practical "how to" information that can immediately be put to use to improve ones machining skills, craftsmanship and productivity.

The purpose of this book is to fill that void and provide concrete suggestions that can help you think and produce like an experienced machinist. If that's what you're looking for, you've hit the mother lode.

This book is primarily directed toward the conventional tool room machinist working in a small shop environment. Many CNC machining suggestions are also included. By virtue of pricing and delivery competition, most small shops have to be very good at what they do. You will find that the equipment and techniques referred to in these pages are commonly seen and used in small machine shops.

Tool room machinists are the ones called upon to do prototype and low production machining. They may also be called upon to build and maintain tools such as dies, molds, and fixtures and occasionally sweep the floor. An accomplished machinist should possess many of the skills of a mechanic, craftsman and problem solver.

Practitioners of the trade are likely to stay quite busy as they face the daily challenges of getting things done. The trade is well suited for results oriented people.

If you're the type of person that frustrates easily then you'd better stay away. If you enjoy a challenge, are mechanically inclined, and have an eye for detail then there are a lot of good reasons to be in the trade. The following are a list of reasons I like it and have stayed with it for so many years.

- You get to destroy as you create. After nearly thirty years working in the trade, I still like making a big mess of chips.

- It is "real" work and you get to produce solid, tangible products you can see and feel.

- There are many different areas of the trade you can go into depending on your preference. Some of the options are: mold making, die making, jig and fixture making, general machining, CNC machining and programming or in some cases you may get to do all of the above. The techniques used in each of these various aspects of the trade overlap nicely and contribute to your pool of knowledge and experience that can be carried over from one job to the next.

- You're in control. Once you become proficient at machining and producing good parts, people will usually leave you alone.

- You get a chance to use your brain once in awhile. The work is not all mundane. You'll have plenty of opportunity to use your brain not only to solve shop math problems but also to solve setup and planning problems. You'll come to appreciate to some extent the schooling you suffered through as a kid.

- The job is somewhat physical but not so much that you will be exhausted at the end of the day. In most cases, you get to move around quite a bit. For my money, it beats sitting behind a desk.

- Jobs are available and abundant. Machinists are needed in every industrialized area of the world. If you don't like where you're working then you can usually find another job without much difficulty.

- Machining is something you can do even as you get older. You won't see many seventy-year-old carpet layers, but you'll see plenty of seventy-year-old machinists and toolmakers still working.

- The "fringe" benefits are nice. I've used company equipment to make hundreds of personal projects.

- The work is relatively safe. You don't hear about many machinists being mortally wounded on the job.

- You don't have to spend a lot of time dressing up and grooming for the job.

- The machines do a lot of the work for you. Once a cut is going, you can relax to some extent.

- You won't have to spend a lot of money to learn to be a machinist. You will however need to spend money for tools.

For the many positives this trade has to offer, there are also a few negatives. The following are things I don't like about this trade.

- Because you are producing solid, tangible parts, it is easy for others to follow your progress and criticize what you're doing.

- The better you are, the more work you'll get. If you are accurate and efficient, you'll end up getting a lot of "hot" jobs and believe me, there will be many of them.

- It can be difficult to erase mistakes. For example, a draftsman or engineer can hit a "delete" button or use an eraser to quickly wipe out a mistake and then proceed from that point. If a machinist puts a hole in the wrong place or cuts a diameter too small he may have to start over. There is no "delete" button for a machinist. This can be costly both financially and psychologically depending on how many parts were made wrong; how much time was spent on each part; and how the critics react.

- Machining is very tool intensive and you can't work efficiently without them. You are constantly looking for, changing, sharpening, buying, making or borrowing tools. Sometimes I'm envious of the computer people that have all their tools right there on the computer screen, just a few clicks away.

- Machinists, for some reason are often treated like second class citizens. One example is air conditioning. You will often find in a company complex, air conditioning every-where except the machine shop.

I could list more but all in all, I believe the positives outweigh the negatives by quite a bit.

I suspect many trades and professions have rules, methods and ideas that get passed along from one generation to the next. In this book I'll attempt to identify and correct some of the misinformation in our trade. One "myth" that comes to mind is that "You should never turn off a surface grinder once the wheel is dressed." The fact is, you can turn the wheel on and off and much as you like without dressing the wheel as long as the wheel is mounted tightly enough so that it doesn't move or shift on the spindle.

No matter how hard we try to avoid making extra work for ourselves, things go wrong. Taps break, cutters break, materials warp, indicators lie, digital readouts skip, milling heads move, vise handles jam, mikes and drawings are misread, parts fall on the floor, get lost, or get made out of the wrong material. It's Murphy of course. (From Murphy's Law) I've spent the better part of my life trying to get one over on that guy. It's not easy, but I've nailed him many thousands of times now and it's always very satisfying.

I'll show you ways of sharpening taps and cutters so they won't break and how to remove taps, screws and cutters that have broken. I'll show you ways to close down over-size holes, minimize material warpage, repair threads and plan jobs to avoid trouble.

Most of the suggestions I've made in this book are techniques and rules of thumb that work for me. It doesn't mean they'll work for you or that you'll even agree with them. I'm always looking for better and easier ways to do things. I'm always experimenting and streamlining techniques that work for me. I suggest you do the same.

To present these suggestions in a concise easy to read format, I've chosen to simply list them.

Some of these rules and suggestions are presented as statements and others are present-ed as questions followed by answers. Some of these rules stand alone, and some of them are followed by explanations and/or discussion.

For you people out there that "know it all" I've included a chapter especially for you. It's called "Tell me something I don't know". In this chapter, I'll present some little known, little understood and in some cases not very useful information about metalworking. For example: How many of you out there know how to make a Slinky®? I mean a real Slinky, not just a curly chip. In this chapter I'll tell you.

In the back of the book, I've included an appendix with drawings of tools that I frequently use in the shop.

It's been said that "metal is man's servant." I've spent many years pursuing that goal so let's get started sharing what I've learned.

chapter 1

Work Fast

As machinists, how often are we asked to produce hardware that was needed "yesterday?" The fact is, quite often as shop personnel try to keep "squeaky wheels" greased. Squeaky wheels come at us from all sides. Production people count on machinists to keep lines going, research and development people count on machinists to keep new product programs on track, maintenance people count on machinists for repair parts and so on.

The bottom line is when people want parts; they want parts. They don't want excuses or anything else. That's one of the beauties of being a machinist. Your responsibilities are clear and simple. If you can get people their blessed parts, they'll go away.

Most people, including myself, don't want to work any harder or faster than we have to. At times though, when the crisis monkey is on us, we have to get the lead out and get going. Crisis machining can be fun once in awhile and generally speaking; any glory to be had usually comes from helping someone through a crisis. If nothing else, it can be a nice change of pace. The hours tend to go by quickly when you're working on a "hot" job.

The following suggestions may help you work quickly. Keep in mind that doing a job correctly has to be your first priority.

1. *Turn man-time into machine-time.*

   This is a popular philosophy for saving time that goes right along with idea that you should never hold back on technology. From a business standpoint the philosophy makes sense but from a machinists standpoint all the CAD/CAM stuff can sometimes take some of the fun out of making chips.

2. *Have lots of tools.*

   This is one of the easiest ways to improve speed. I like to have and use my own tools even if the shop is well stocked. Shop tools are never put back in exactly the same place and you never know what condition they'll be in when you find them.

   If you're just starting out in the trade, I recommend buying a set of tools from a retired machinist. That way you'll hit the ground running and you'll immediately be worth more to your company. Machinist's tools hold their value well so be prepared to pay a fair price. You can upgrade later as you see fit.

   Avoid borrowing tools over and over, if you use a tool often enough then either buy one or make one.

   Don't waste your time making easily purchased tools like 1-2-3 blocks and V-blocks. Instead, spend your time making custom tools and fixtures that aren't easily purchased that are tailored to the machines you use and the type of work you do.

3. *Use dedicated tools.*

   This business of having one tool that does everything isn't very efficient. You often see these kinds of tools advertised in the back of the mechanic's magazines. For the most part, each tool you have whether it is a hand tool or power tool should serve just one purpose. You want the tool to be ready when you need it.

   A simple example would be a screwdriver. You can buy a screwdriver with interchangeable tips; the idea being that with one screwdriver handle you can cover all of your screwdriver needs. I suppose that is true if you want to fiddle around with the tips, changing them, dropping them, and losing them. It's faster to grab a tool that is ready to go.

   Another example would be having an assortment of dedicated air tools. I have a drawer full of cheap air spindles; each mounted with different cutters or abrasives. If I need to cut off a pin for example, I can be cutting within seconds, instead of fiddling around with wrenches, collets and Murphy's little surprises.

FIGURE 1–1   Full width cuts are the fastest way to clean up a surface and also make parts look better.

I suppose you could get carried away this type of thinking. I wouldn't go so far as to buy a handle for each socket I have, nevertheless, having and using dedicated tools can greatly improve your speed.

4. *Make all your parts the same.*

   This is a great way to expedite just about any job. If you're making multiple parts, by making all the parts the same you'll always know exactly where you stand with the job.

   If you have parts all over the map dimensionally you'll constantly be re-measuring, dealing with "special cases" and hassling to get consistency and control over a job.

5. *In a milling machine, when practical, use a large enough cutter to cut across the entire surface of the part in one pass. (See Fig. 1-1)*

   This is an effective way to save time and also make parts look better.

   Taking several passes with a small diameter cutter to clean up a surface is usually a waste of time. It is much faster to cut an entire surface in one pass when practical.

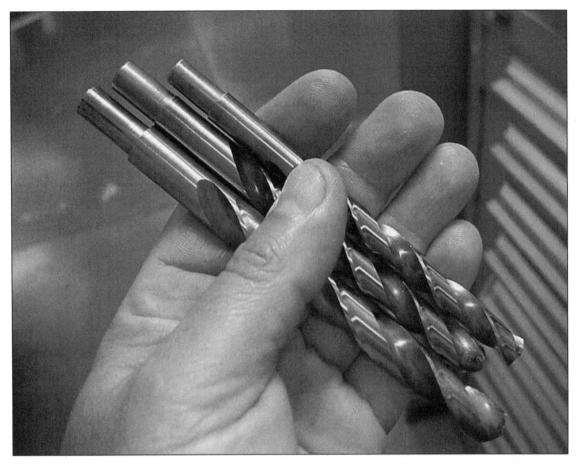

FIGURE 1–2   The shanks on these drill bits have been turned down to common collet sizes so they can be mounted in the spindle without having to use a drill chuck.

6. *Turn the shanks of your larger drill bits down to common collet sizes. (See Fig. 1-2)*

   I dislike cranking the knee of my mill up and down to accommodate a drill chuck. If you turn the shanks of your larger drill bits to a common collet size, you can avoid that hassle. You'll be able to use standard size collets to hold your modified drills, without having to use a drill chuck. You can also do this with reamers and other cutters.

7. *Use stub drills. (See Fig. 1-3)*

   Anytime you can drill a hole without first center drilling you simply save whatever time it would have taken to do that center drilling. Normally, a high percentage of holes in parts are simply clearance holes used for bolting parts together. Clearance holes are usually anywhere from .015″ to .030″ larger than bolt diameter.

   If you know you're going to be drilling clearance holes or other non-critical holes then you can use stub drills without center drilling. A stub drill that has been properly web thinned (See Fig. 7-4) will cut with little pressure and will produce a surprisingly accu-

FIGURE 1–3   Stub drills can be used to drill holes without first center drilling.

rate hole. You can either buy stub drills already made or you can make them by cutting off standard length drill bits and regrinding the tips.

Even if the stub drill runs out a little bit as you start a hole, you'll probably have enough tolerance on a clearance hole so that it won't matter. If the hole is deep, or has to be precisely located then it is best to center drill first to maintain accurate location.

8. *Use a speed chuck in a conventional milling machine. (See Fig. 1-4)*

You can change bits with these chucks without turning off the spindle. They work great for quickly changing from a center drill to a drill, which is one of the most common tasks performed on a milling machine.

9. *Use a slide fixture in a conventional mill for drilling holes. (See Fig. 1-5)*

A slide fixture saves vise clamping time. That may not sound like much but if you have many parts to drill then the savings become apparent. One way to make this setup is to close down the vise on something that is a few thousandths wider than the

FIGURE 1–4   You can change cutters in quick change drill chucks without stopping the spindle. This chuck was made by Wahlstrom Chuck Co.

parts. That way the parts will locate accurately yet still slide in and out of the jaws easily. Gauge blocks work well for this purpose. As an added benefit the gauge blocks hold the parallels in place. You'll also need to set up some kind of stop for locating the parts. Using stub drills in combination with a slide fixture is a great way to make good time on conventional drilling jobs.

10. *Use chip color to determine speed, feed and depth of cut in ferrous materials.*

Most machinists in small shops using conventional equipment set feeds, speeds and depths of cut based on feel and experience. With a little practice, a newcomer can soon get the hang of it.

The best rule of thumb regarding this subject is the tan chip rule which is as follows: If you want your cutter to last, then your chips should come off the workpiece no darker than a light tan in color using high-speed steel or cobalt cutters and brown using carbide cutters.

FIGURE 1–5   A slide fixture allows parts to be changed without loosening and tightening the vise.

If you push a cut much beyond that point your cutter will almost certainly start to break down. Once a cutter begins to dull, the resulting heat and friction tend to accelerate the breakdown.

Cobalt and high-speed behave a little differently than carbide as to when and how they break down.

Cobalt and high-speed hold an edge very well up to a certain point. If that point is exceeded then the edge quickly breaks down.

With cobalt and high-speed that "point" is determined by the combination of speed, feed, and depth of cut that gives you a light tan chip. Once you've found a combination that gives you a light tan chip you can make adjustments from that point depending on what you're doing. If you're roughing, you'll probably use a faster feed and slower spindle speed to maintain the tan chip color. If you're finishing, then it'll be the other way around.

Carbide, on the other hand, tends to break down more gradually than high-speed or cobalt. In other words, carbide doesn't have the abrupt point of failure like the others do.

Though the "tan chip rule" works fairly well as a yardstick for measuring the aggressiveness of a cut, there is an exception. Chips that turn blue some distance after they leave the cutting tool are usually not detrimental to the cutting tool.

Having watched chips come off stock under various conditions for many years, I've come to the conclusion that there are two separate sources of heat generated while cutting metal in a machine tool.

The first source of heat is the result of friction from the metal moving across the cutting edge. The second source of heat results from the metal in the chip being deformed or upset as it is forced to flow across the cutting tool.

This second source is what causes chips to turn blue after they are already some distance from the tool. The heat generated from the deformation literally takes that split second to migrate to the surface of the chip.

11. *Rough ugly.*

I believe there is some truth to the idea that "roughing is where you make your money." Bear in mind that you can't do much roughing if there is little material to rough off. The most efficient roughing takes place in a saw. Within reason you should try to remove as much material as possible or practical with a saw.

Roughing is sort of a behind the scenes operation where you get to do it as fast and ugly as you want. Roughing is one operation where you get a chance to erase your tracks later on. Take advantage of the situation.

12. *Work your way up to a heavy roughing cut.*

When I'm searching for an aggressive cut in a conventional machine, I like to feed the tool by hand first before I use the auto feed so that I can "feel" the cut. If you encounter an excessive amount of noise, backpressure, vibration or resistance when test feeding then you may have to adjust the cut somewhat.

Note that by locking the knee of your Bridgeport, you can increase the rigidity of the system which in turn allows for more aggressive roughing.

13. *Avoid using a single flute fly cutter to rough with.*

A single flute fly cutter is best used as a finishing tool. You're better off from the standpoint of quickly removing stock to use a multi-tooth cutter of some sort like a corn-cob cutter or a multi-tooth insert cutter. In my opinion it's hard to beat a cobalt corncob cutter for heavy roughing because of the abuse they can take. I like the round insert face cutters for light roughing and finishing because they hold up well and when the inserts get dull, they can be rotated to expose fresh cutting edges. (See Fig. 1-6)

FIGURE 1-6   Face milling can be accomplished with various types of cutters. The insert cutter on the left uses round inserts which can be rotated to expose fresh cutting edges. For rough milling it's hard to beat a short, beefy corncob type cutter.

14. *Try to rough as close to final size as practical.*

    For finishing tools to stay sharp you want to avoid working them too hard. In a conventional machine, leaving .010″ to .030″ stock for finishing works well. In a CNC machine you may get away with leaving less because of the machine's consistent accuracy.

15. *Work your machine hard when roughing, but do it the right way.*

    You want to make your machine groan, not beg for mercy.

    I believe increasing your feed is generally the best way to remove stock quickly. By keeping depth of cut and spindle speeds moderate you may be able to increase feed to get things moving.

    Increasing depth of cut also works but that puts a lot of pressure on the cutter and components of the machine. What you want to do is put load on the motor. If you can hear the motor bog down when a large diameter cutter enters the material then you can be confident you're working the machine hard without abusing it.

16. *Place your hand on a milling machine table to gauge the pressure of a cut.*

    Placing your hand on a machine table when the machine is cutting allows you to feel how much the table is deflecting under load. If you can only mildly feel the cut through the table then the machine is likely working below its capacity. This test also works on CNC milling machines where it is sometimes difficult to gauge the amount of pressure you're putting on the machine or cutter.

17. *Make parts with as few setups as possible.*

One way to do that conventionally is to do your finishing right after you rough. To make good time, ideally you would like to rough in a surface then immediately finish that surface without removing the part or changing cutters. Then after you finish the surface, you can break the setup to prepare for the next cut. That way you don't have to repeat setups.

What you would like to avoid, if possible, is roughing in all the surfaces then going back and finishing all the surfaces. If you do that you'll be making many of your moves and setups twice.

Unfortunately it can't always be done. On parts that warp easily such as thin or hogged out parts, you may have to rough in all the surfaces first. That way when you go back to finish the part, you can cut out any warp that may have occurred during the roughing operations.

18. *Use air mist to prolong the life of your cutter and increase stock removal rates.*
    (See Fig. 1-7)

A little air/water mist helps cool and preserve cutters. A lot of machinists use air/mist sprayers. I don't use them much simply because I find them to be too much hassle. If you follow the tan chip rule mentioned previously, a mist sprayer is not necessary.

FIGURE 1–7   A mist sprayer is used to keep the end mill cool

19. *Go as fast as you dare in aluminum and other easily machined materials.*

It is difficult to wear out a cutter in aluminum especially with a conventional machine.

Machining aluminum is sort of like driving on the autobahn. You can basically go as fast as safety and common sense dictate. You don't want to go so fast that you end up crashing. But when conditions warrant, such as when you have a rigid setup and you have a lot of stock to remove then you can put the pedal to the metal.

You can even use the rapid traverse feature on your conventional mill to increase feed rate. The rapid traverse rate on the mill I use is somewhere around sixty inches a minute. Make sure you run the spindle fast enough to maintain a reasonable chip load somewhere in the .010″ to .015″ range.

In a CNC machine you can literally fly through aluminum if your setup is rigid and you have a lot of stock to remove. I've used a five hundred inch per minute feed rate with a one inch diameter three flute end mill turning at 10,000 RPM. That's as fast as the machine would go. If you do the math you'll see that those parameters produce a chip load of about .016″. Bear in mind there is no reason to run that fast if there is little stock to remove and your cuts are short.

If you're not using a fast feed then there is no advantage to running a spindle at warp speed. In fact, it is usually a disadvantage because there is a tendency for things to start chattering when spindle speeds are too high.

20.   *Bore holes with a mill like you would with a lathe.*

For some reason many machinists use very slow spindle speeds when boring holes with a boring head. They look like they're stirring taffy.

There is usually no need to run a boring head at such slow spindle speeds unless the boring bar is flimsy and prone to chatter. You can usually use the "tan chip" rule for setting feeds and speeds, just like you would if you were boring a part in a lathe.

Use short, stout boring bars when you can. A short, stout boring bar will help eliminate chatter and won't spring away as easily as a long thin boring bar. It is difficult to hold size with a thin, springy boring bar.

As far as I'm concerned, you can never have too many boring bars. You'll need a variety of different size boring bars to match the different hole diameters and depths you'll encounter.

21.   *Power tap blind holes.*

Do it under the right conditions or chances are you'll be digging out a broken tap. Many instructors would just say "never power tap blind holes."

If you drill a tap size hole at least twice as deep as the threads you need, chips will have a place to go and won't cause binding. Use a spiral point or gun tap so chips get pushed ahead of the tap. A spiral point or gun tap won't bind like a plug tap or hand tap. As long as you use a sharp tap with some cutting oil and give the chips a place to go, the tap should cut freely.

If the design of the part is such that you don't have enough material to drill a hole at least twice as deep as the thread then it is safer to either hand tap to final depth or use a tap that pulls chips out the top. By hand tapping, you can gauge the amount of torque you put on the tap and you can clean the chips out as you go.

The spiral fluted taps that pull chips out of the tops of holes are weaker than other taps and are best used in aluminum and other easily machined materials.

22. *Saw your raw stock about a tenth of an inch larger than finished size.*

It is certainly possible to saw closer than .1″ but for me it's not worth the extra effort.

Most saws are not high precision machines and often blades are in less than ideal condition. A tenth of an inch gives you a little breathing room if the saw blade runs off a little bit or if your stock isn't held square to the blade for some reason.

If I have many small parts to make, I'll try to saw a little closer than .1″, maybe to within about 1/16″ of finish size. You can usually get away with cutting small parts closer because the saw blade has less chance to run off.

FIGURE 1–8　Cold cut saws work great for cutting off bar stock. This machine was made by Doringer Mfg. Co.

23. *Use a cold cut saw for cutting off bar stock. (See Fig. 1-8)*

You can really make hay with these saws. They are especially useful for cutting off bar stock. For some reason these saws have not gained the popularity they deserve probably because the people in charge of purchasing equipment are not familiar with them. Some of the advantages they have over horizontal band saws are that the blade is rigid and durable which enables them to cut stock cleanly and close to size. They are also compact and easy to operate.

Although similar in design, these saws work on a different principal than abrasive cut-off saws. Cold cut saw blades turn relatively slowly and cut heavy chip loads unlike abrasive cutoff saws.

24. *Do one operation at a time in a tool room lathe when running multiple parts.*

This suggestion applies mostly to small parts that are held in collets.

FIGURE 1–9   Stack milling is an effective way to save time. The trade-off is that setups take more time and tolerances are usually more difficult to hold.

Don't try to run a tool room lathe like a turret lathe. It's faster and easier to change parts with the collet closer doing one operation at a time than it is to change tools and settings.

25.   *Change small lathe parts when using a collet closer without turning off the spindle.*

26.   *Stack parts when you can. (See Fig. 1-9)*

Stacking parts and machining them all at once can save a substantial amount of time. It can be a double-edged sword however. It is more difficult to hold tight tolerances on stacked parts. Also setups are usually more involved and time consuming when stacking parts.

If you have to move a stack in order to do other machining operations, try to keep the parts clamped together so they move as one block. In other words, try to avoid moving the parts in relation to one another.

27.   *Stack parts on edge for drilling and tapping. (See Fig. 1-10)*

Drilling and tapping a stack of parts on edge can save you time but you have to be careful how you do it so that you don't accumulate error.

Let's say you have a stack of twenty 1/4″ plates that need to be drilled and tapped on their edges. After clamping the stack in the vise, you have to measure the width of the stack and divide that measurement by twenty to get the distance you should move between holes. Let's say your overall stack measures 4.94″. The real distance between the centers of the plates is 4.94 divided by twenty or .247″. That's how far you would have to move each time to keep the holes centered in the plates. If you moved over .250″ each time, your holes would become increasingly off center as you advanced.

FIGURE 1–10 A stack of parts is drilled and tapped on edge.

28. *Consider buying pre-squared blocks for high quantity runs.*

There are vendors that specialize in supplying material squared to size so that you don't have to do the squaring. Having blocks prepared this way is a great way to get a job off to a fast start.

29. *Use an end mill in the lathe to rough out a flat bottom hole. (See Fig. 1-11)*

Using an end mill in a conventional or CNC lathe allows you to rough out a flat bottom hole very close to depth with a square corner. Any subsequent boring you do to finish the hole will take less cutting than if you used a standard drill bit to rough out the hole. You can hold an end mill in the tailstock chuck.

An end mill fed in with the tailstock will usually cut a hole a little larger than the end mill diameter so be careful that you don't cut an oversize hole. An end mill used this way will cut with less pressure if you drill a pilot hole first. Use the tan chip rule for setting spindle speed.

FIGURE 1–11 An end mill is used in the tailstock to rough out a counter bore.

30. *Keep a box of loose drill bits handy.*

    If you need to drill an approximate size hole such as a pilot hole for example, there is nothing faster than just grabbing and using an approximate size drill from a box of loose bits.

31. *Keep a large assortment of T-nuts on hand.*

    Angle plates, rotary tables, drill presses and other machines and fixtures around the shop are going to have different T-nut slot sizes. Not having the right size T-nut for a piece of tooling can be frustrating when you want to keep a job moving. T-nuts are cheap. Open a catalog and buy yourself a large assortment of T-nuts to avoid getting hung up.

32. *Avoid clearing your cutter to go back for another cut.*

    If you are roughing off material in steps in a conventional mill or lathe, don't bother clearing the tool to return for the next cut. If the roughing tool rubs or grooves the

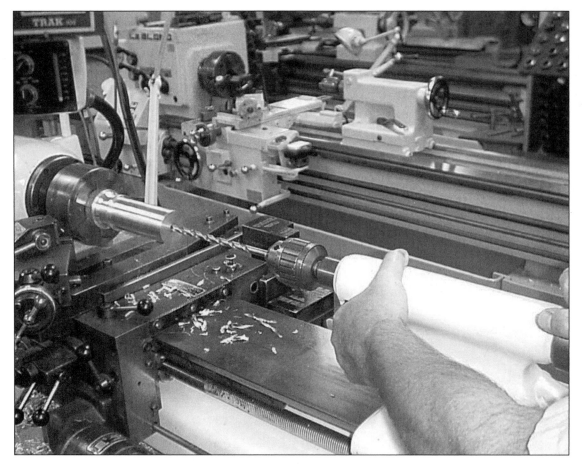

Figure 1–12 A deep hole can be drilled with a conventional lathe by sliding the tailstock.

part as it is returned to the starting point, so be it. This technique saves you from having to look at your dial settings all the time.

Don't, however, take finishing cuts over a rough surface with swirls or grooves. Clear the tool when you are within finishing range so the tool cuts with even pressure over a consistent surface for finishing.

33. *Don't bother removing a built up edge on a lathe tool when roughing.*

Your cutting tools may accumulate a built up edge when roughing. A built up edge or BUE is a slight accumulation of material from the workpiece that adheres to the cutting edge of your tool. A tool will still cut with a built up edge, at least well enough to rough with.

To make smooth, precise, finishing cuts it is best to remove the built up edge so that chips slide off the tool freely and the edge of the tool does the cutting instead of the built up edge.

Removing it is often easier said than done. On carbide, if you attempt to just pick it off with a knife or something, you'll likely pull off a bit of the carbide and ruin the edge. If you attempt to stone or file it off you may end up rolling the edge over somewhat which is also no good.

You can remove a built up edge by applying a liberal amount of cutting oil to the tool and then abruptly but intermittently hand feeding the tool into some stock. The built up edge will usually get pushed off by the resultant chip.

34. *Drill deep holes in a conventional lathe quickly by sliding the tailstock. (See Fig. 1-12)*

Chip packing becomes an issue when drilling deep holes. The deeper you drill the harder it is for chips to get out. Instead of winding the drill bit in and out with the tailstock crank you can manually push the tailstock in and out along the ways of the lathe. When the drill bit packs, loosen the tailstock and pull it back to free the chips. Then push it back in until the drill bit bottoms. Re-tighten the tailstock and drill a little more using the crank. Repeat the process until you are to depth. You'll be surprised how fast you can drill a deep hole with a conventional lathe using this method.

Always begin drilling deep holes with a standard length drill to get the hole started straight. Drill as deep as you can with a standard length drill before switching to a longer one. This suggestion applies to deep holes drilled with a lathe and a mill.

35. *Deburr rough edges with a small, angle-head air spindle. (See Fig. 1-13)*

Burrs and therefore the task of deburring are the unfortunate by-products of machining. There is no way to completely avoid throwing burrs when machining.

After rough cutting stock to size, you usually have to remove the resultant burr to begin machining. You can remove those and other fairly large burrs by sanding them with a small sanding disk mounted in an angle head spindle. The advantage of this method is that you can do the deburring at your machine between cuts instead of having to file or walk to a disk sander.

There are a few things you can do to minimize burrs and the effort it takes to remove them. One method you can use is to climb mill into material. The use of sharp cutters also minimizes burrs.

36. *To save a little time, avoid turning off a conventional milling machine to change parts.*

Some common sense is needed here. If the setup is such that you can move the cutting tool beyond the part by at least a few inches then you can usually change parts safely without turning off the spindle. If you are going to measure the part, file the part or do any operation other than just changing or removing the part then it is best to turn the spindle off.

FIGURE 1–13 An angle head air spindle can be used to quickly remove heavy burrs.

Be especially careful around cutters that are rotating close to another piece of solid tooling like a vise or something. That is potentially a more dangerous situation than having a cutter that is just rotating out in the open.

I know of two accidents that happened as a result of the machinist not staying clear of a cutter that was rotating next to a solid piece of tooling. One machinist was using a milling machine to cut screw slots in some small parts with a slitting saw. When he went to change a part, he got his fingers caught in the small opening between the rotating saw blade and the holding fixture. The blade sucked three of his fingers through the small opening. Fortunately the opening was big enough that he didn't completely chop off his fingers.

The other incident happened in a surface grinder. The rotating grinding wheel was positioned about a quarter of an inch above the magnetic vise when the machinist decided to use his hand to wipe some debris off the magnet. Four of his fingers were ground down as they passed through the opening between the wheel and magnet.

FIGURE 1–14 An arc is being machined around the end of a small part by rotating the part on a dowel pin.

37. *Cut the diameter of a lathe part instead of the face to remove material quickly.*

    If you have a lot of material to remove from the length of a lathe part, it is faster to remove material by taking cuts off the diameter rather than the face of the part. Try it, you'll see.

38. *Cut arcs by hand in a conventional milling machine. (See Fig. 1-14)*

    To cut an arc around the end of a part, use a dowel pin as a center pivot and rotate the part around by hand. Mount the dowel pin in your milling machine vise with a V-block. I've used this method several times with good results. Be careful. Begin by cutting the protruding corners of the part first so that the cutter does not suddenly jerk the bar and break the cutter.

FIGURE 1–15 These tools make cutting spherical shapes a breeze with a conventional lathe. This tool was made by Holdridge Mfg. Co.

39. *Cut spherical shapes with a conventional lathe. (See Fig. 1-15)*

These tools are easy to set up and work well for quickly and consistently cutting spherical shapes conventionally.

40. *Use short, stubby end mills whenever possible. (See Fig. 1-16)*

Short, stubby end mills don't deflect as much as longer end mills and as a result last longer. End mills with flute lengths about one and a half times the diameter of the end mill or less, don't flex much and can be pushed harder than longer end mills. It's hard to beat a short, stubby corncob type rougher for removing material quickly.

41. *Avoid pushing small end mills too hard.*

If you have to use a small diameter end mill, be prepared to take your time. End mills under an eighth inch in diameter simply can't be pushed very hard. They'll break if you do.

FIGURE 1–16 Stubby end mills are rigid and should be used whenever possible.

I've found that when it comes to milling with small diameter end mills, a six sided single flute cutter can withstand more side pressure than a multi-flute helical end mill of the same diameter. Hex cutters can be quickly made on a surface grinder or cutter grinder. (See Fig. 1-17)

When making carbide cutters like this, it is best to lightly break the vertex of the hex on the opposite side of the cutting edge to reduce the tendency for the cutter to chip out. You can use either a diamond file or diamond grinder to break the edge as shown in the right hand photo.

42. *Measure stock one time only to rough in a feature.*

Some machinists waste time by re-measuring after every roughing cut. It's not necessary. Measure stock carefully one time before you begin roughing to determine how much stock you need to remove, then don't measure again until you are within finishing range. Either that or cut to a scribe line.

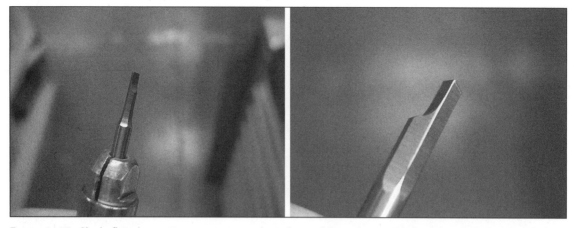

FIGURE 1–17 Single flute hex cutters are easy to make and can withstand more side load than helical end mills. They work great for cutting O-ring grooves and other features requiring deep, narrow cuts.

43. *Avoid tilting the head of a mill when possible.*

When cutting or drilling an angle in a part, it's best to find a way to either tilt the part or use an angle cutter to do the job. It is awkward to use a mill with a tilted head. Furthermore, once you tilt the head, you're obliged to tram it back in.

44. *Take your drawing with you to the stock room.*

When searching through stacks of material in the stock room, you don't want to have to remember what you are looking for. It can be a little confusing at times especially when you don't have exactly the stock you need. Save yourself a trip. Take your drawing with you to the stock room.

45. *Use a high volume air nozzle.*

This can be a controversial subject. In some shops, the rule is you are not supposed to use air at all to blow chips off machines. Some people say that cleaning a machine with compressed air ruins the machine. They believe the air stream forces small chips and debris between the table and ways of the machine and causes binding, scratching and rapid wear on the ways.

A little common sense is needed here. If you are going to use air to blow off chips, don't aim the air stream directly into the junction of the machine table and ways. Aim or blow the chips away from areas of the machine that may trap chips.

Adjust your cleaning procedure according to the type of material you are cutting. If you are cutting aluminum or other soft, "clean" materials, you probably won't damage anything by using air. On the other hand, if you are cutting through hard, gritty or abrasive material, then it makes sense to use a brush and paper towel to wipe the grit off the ways of the machine. Also, if you use sandpaper or other abrasives in or around your machine, make sure you wipe the grit off with a paper towel or rag before you move the table.

With that in mind, get yourself an air nozzle that puts out some volume so you can blow chips off quickly.

46. *Look at "numbers" once in a while.*

On conventional machines I generally set cuts based on feel and intuition. Later I may run some numbers to get a quantitative idea of what is going on.

Two important numbers to look at when it comes to machining are "chip load" calculated in thousandths per tooth or revolution and "Surface feet per minute" which is a measure of the rate at which material slides across a cutting edge.

The formula for chip load in a milling machine is:

Chip load = Feed Rate / RPM X No. of Flutes

The formula for SFM is:

SFM = RPM X Dia. X .262

For a more in depth discussion on using these formulas see Chapter 15 on CNC machining.

47. *Tidy up your desk or bench by handling objects one time only.*

This suggestion sounds simplistic but is effective. When tidying up or putting things away, there is a human tendency to pick something up and set it down again somewhere else. Avoid that tendency. Whatever you pick up to put away should be put away the first time you handle the object.

48. *Make notes about machines you don't use often.*

Machines you don't use to often may take some time to get used to after you've been off them for a while. Make written notes to yourself on the ins and outs of any particular machine so you'll be able to get back up to speed quickly.

Conclusion

As you can see, there are various approaches to running jobs and setting cuts. Ideally, machinists strive for the most aggressive cut they can get away without quickly ruining cutters or trashing setups. Often, the limiting factor in terms of aggressive cutting turns out to be the rigidity of the setup and the strength of the cutter. An additional factor to consider is that parts become weaker as they're machined.

Machinists have to be intuitive stress engineers. They constantly have to make judgments about the aggressiveness of cuts based on the type of material they're cutting, the rigidity of the machines, the rigidity of setups and the strength of cutting tools. It's a skill that comes with practice.

It is unlikely any two machinists would run a job the same way or at the same rate. Most machinists wouldn't run a job themselves the same way each time. There are simply too many variables and too many different ways of doing things.

Occasionally you'll hear a machinist say "Why should I work fast? I'm in no hurry. I get paid the same hourly rate regardless of how fast I work." That may be true, but by learning how to work fast you'll have that option when you need it.

chapter 2 Get It Right

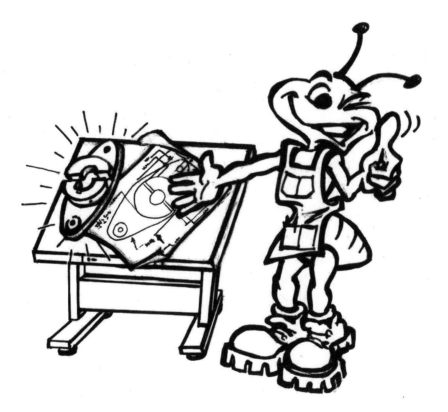

One of my favorite quotes as it relates to working in a machine shop was spoken by a veteran toolmaker I once worked for. It was in response to my complaining about having wasted time working on a part that was already out of tolerance. He said: "All I know is if I don't check it…it's wrong."

It seems you can double check things all day long without finding an error then the first time you don't check something that's the measurement that'll be wrong.

Part of learning any trade is learning how to avoid trouble. Mistakes are no fun. The trick is to try to develop habits and ways of working that reduce your chances of making mistakes without wasting time.

The following is a list of suggestions that may help you stay out of trouble. Everybody is different in how they approach jobs and each of us has our own strengths and weaknesses.

With that in mind, some of the following suggestions should give you some insight into avoiding errors.

1. *There is no such thing as a simple job.*

    I've seen so many simple jobs screwed up, I've come to the conclusion there is no such thing. So called "simple" jobs get screwed up because people have a tendency to start cutting without planning or double checking anything. You have to train yourself to identify "simple" jobs and be extra careful on them. Believe me, it's easier said than done.

2. *Measure raw stock before you begin machining.*

    I've seen a lot of grief caused as a result of people not measuring or double checking raw stock size. Before machining begins, and that includes saw cutting, you should always check raw stock size to make sure it is what you want.

3. *Layout work for consistent results.*

    This suggestion applies mainly to parts that are going to be machined conventionally. You won't see many parts laid out for CNC machining simply because the layout, in essence, is done in the CAD (Computer Aided Design) system.

    Some people think scribing layout lines on parts is a waste of time. In most cases I believe it is time well spent when conventional machining. Not only do layout lines drastically reduce bad cuts; they also make parts easier and faster to machine. After squaring material to size, I almost always layout hole locations, pockets and other features on the first part.

    Laying out work is also a great way to check drawings for errors or omissions. If the drawing provides enough information to make the layout then you can usually make the part.

    You can use calipers to make layout lines. When I first started machining, I used to be shy about scribing layout lines with calipers. I'd look around to make sure nobody was watching and then quickly use them to scribe a line. I've come to the conclusion that using calipers is one of the fastest and easiest ways to scribe lines. I don't expect lines scribed with calipers to be too accurate. I use them mainly as a double check. You can mark areas for scribing with a black felt tip pen beforehand to make the layout lines stand out better. The ink from a felt tip pen is easily removed with rubbing alcohol.

    If a part is complex and needs many lines, I may use a height gauge to make the layout. You can get a cleaner, more accurate layout with a height gauge than you can with calipers.

On lathe parts, you don't have to scribe layout lines. Instead, you can use the cutting tool when the part is spinning to create a target mark to roughly cut to. That way you don't have to constantly watch dials and readouts.

The use of layout lines and visual targets makes machining easy.

4. *Lightly spot the surface of a workpiece before drilling holes.*

    Another way to double check hole locations is to first lightly spot the surface of the work with a center drill then check the location of the spot with a scale. If you happen to be off for some reason, you'll be able to correct the error before drilling in the wrong location and possibly ruining the work.

5. *Double-check measurements and calculations.*

    This is an important rule for making good parts. We all make mistakes during the course of a day, that's a given. Some of us make more mistakes than others; therefore we have to try to catch our mistakes before we make a bad cut. The best way to do that is to double check, preferably with two different methods. When you double-check, you give yourself a huge mathematical advantage. You'll probably catch any mistakes before you make a bad cut.

    To illustrate my point, let's say you make one mistake for every hundred calculations or measurements you make. Making cuts based on that ratio would produce an excessive amount of errors. If you double-check yourself, in theory you would reduce your chances of making an error to "one in ten thousand." Use different methods to double check to insure that you don't make the same mistake twice.

    Double-checking becomes second nature after awhile and is often rewarding. A mistake isn't really a mistake until you've made a bad cut. A similar analogy in the construction trade would be: "It ain't screwed up 'till the cement's been poured."

6. *Tighten cutters sufficiently.*

    One reason parts get messed up is because the cutter comes loose in the tool holder. This happens mostly during roughing cuts or when fly cutting because of the increased pressure and vibration on the cutter. Avoid grief; get a good grip on the cutter before you begin.

7. *Get a good grip on a workpiece.*

    This is sometimes easier said than done. There are a few things you can do to keep parts from shifting.

    • Err on the side of using too many clamps in your setups.

- Use the center of your vise whenever you can to hold parts securely.

- Clamp as close to a cut as possible.

- On curved and difficult to hold parts, build a nest of some sort to support the part.

- Make sure tie down studs are as close to the part as possible.

8. *Clean and stone set up blocks, parallels, and machine tables often to remove dirt, chips, craters and burrs. (See Fig. 2-1)*

This suggestion is more important for complicated and time-consuming setups.

It can be aggravating to have to tear down a setup because a chip or nick skewed the setup. Build a good, clean setup the first time to avoid wasted effort.

9. *Set work stops close to the area you'll be machining.*

This suggestion is more relevant for parts that protrude well above the vise or table. Since no block is perfectly square, setting a stop higher up the part will locate the upper end of the part more accurately.

10. *Be aware that the head of your mill may not be trammed in.*

Many instructors would say "Always check the head of your mill for perpendicularity or tram before you start using it." Most of us don't check anything on our machines before we start using them. We just turn 'em on and go. The next best thing then is to be on the look out for an out-of-square situation such as non-parallel surfaces.

I've seen more grief caused by the head of a mill not being trammed in than just about any other situation. It's a perfect recipe for bad feelings. The out-of-tram confrontation usually goes something like this: One guy says "Was that you that tilted the head on this mill?" The other guy replies "Yeah, what about it?" The first guy responds "Well, I screwed up this part because you didn't tram the head in." The second guy says "Well, didn't you check it? Do I have to do everything for you?" The first guy says, "You could have at least told me or left me a note or somethin'." The second guy says, "Nobody ever left me nothin', I look out for myself." The first guy says "I just lost two days worth of work because of you." And so on.

You have to be especially careful if other people use your mill. If you're planning on doing some precision machining then you had better check the tram.

11. *Tram the head of your mill with an angle plate. (See Fig. 2-2)*

The quickest way I've found to check the tram of a vertical mill is to simply clamp a precision angle plate in the vise then sweep the top of the angle plate with an indicator. That way you can quickly determine if the head is square or not. It's not the most

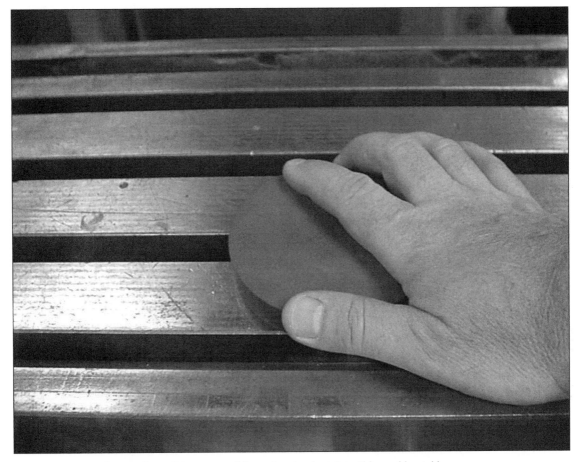

FIGURE 2-1   A soft stone is used to remove nicks and protrusions from this machine table.

FIGURE 2-2   The head of the mill is being checked for tram or perpendicularity. The most precise method is shown using the 1-2-3 block.

precise way to do a check, because of the stack up of tooling involved, (the vise and the angle plate) and the small indicator sweep area, but it is fast. You will at least catch any glaring out-of-tram errors. Then you can decide how to proceed. The most accurate way is to remove the vise and sweep a wide area of the table with an indicator. Make sure you lock the knee of the machine before you begin tramming the head.

Much "out-of-tram" grief could be avoided if people would follow the next suggestion.

12. *Never leave a machine close to being correctly lined up.*

There are other, less hideous ways of fouling up your fellow man.

If you are going to leave the head of a milling machine tilted then leave it at a huge angle so that it is obvious. Or put a note on the machine stating that it is out-of-tram.

The same goes for the milling machine vise. If you are not going to lineup the vise precisely, then leave it way out of square, or put a note on it stating that it is not square.

The same goes for the lathe tailstock. This is something that is not often moved.

If you do move it, either indicate it precisely back on center or leave a note stating that it's not on center.

If you leave a setup or something close to correct, the next guy that comes along is going to assume that it is exactly correct. That's when the problems start.

13. *Tram the head of your mill after heavy roughing cuts.*

14. *Avoid tearing down setups until absolutely necessary.*

It's amazing how often we return to a set up. Sometimes you have to make a part over, modify it or tune it up in some way. Leave your setups intact as long as possible. You'll be glad you did.

15. *Avoid unchucking complex parts too quickly.*

A final check before unchucking a complex part is usually time well spent. If possible, take a little break before removing the part. You'll be in a better state of mind for doing a final check.

16. *Plan for mind drift.*

You've probably heard it said many times: "Keep your mind on the job" It's easier said than done and in my opinion not very realistic. There are a million reasons why our minds drift.

If you can force yourself at the beginning of the job to double-check your calculations and lay out your work then you can sometimes get away with total mind drift. It's risky though.

17. *Avoid quick glances at drawings.*

    There is a direct correlation: The quicker the glance, the higher the risk of picking up a wrong dimension.

18. *Avoid finishing a part until you've roughed it in.*

    This doesn't apply to all machined parts so you have to use your judgment. If you are removing a large percentage of stock to arrive at the finished part then it is best to leave a little material for finishing especially on the important features of the part.

    A few reasons for this are: First, the part may warp as material is cut away. Second, the part may move in your holding fixture from the pressure and vibration of the roughing cuts. Third, the part may expand or distort from the heat generated during roughing.

    Once most of the raw stock has been removed and the part has cooled then light-finishing cuts can be made confidently and accurately on a stable "non-moving" piece of metal.

19. *Avoid pushing reamers hard against the bottom of blind holes.*

    A reamer will cut oversize if you do.

20. *Avoid jamming a vise handle into the ways of the mill. (See Fig. 2-3)*

    It is safer if a vise handle can't reach the ways of the mill. You may have to either cut down your existing handle or purchase a "ship's wheel" handle.

21. *Make a habit of cranking machine handles in the same direction when approaching and stopping on dial settings.*

    By doing so, you insure against backlash error. I've always made a habit of cranking machine handles clockwise to approach dial settings including the knee crank on a milling machine.

    Be aware that your lathe compound also has backlash in the lead screw and may move. Keep an eye on the direction of the compound backlash since it may be important depending on whether you're boring an ID or turning an OD. Tighten the gibs of the compound when not in use so that the compound doesn't move easily.

FIGURE 2–3   The swivel vise handle is shown jammed into the ways of the mill...a condition that must be avoided.

22. *Back off and return to dial settings.*

Vibration and shock cause machine tables to drift. For example, in a lathe the sudden shock of a collet closer closing will often cause the cross slide to move. To maintain dimensional repeatability, it is best to back off and return to dial settings each time you engage the collet closer. Hammering parts in a vise or chuck can also cause tables to move. Backing off and returning to dial settings is a good way to help maintain dimensional accuracy and repeatability.

23. *Remove handle pressure to keep machine tables from drifting off location.*

In a conventional mill, the best way to keep the table from moving off location is to back the pressure off the handles once you reach your dial setting so that there is no preload whatsoever on the lead screw. Once the handles have been backed off, the table clamps can be tightened with little risk of the table moving.

24. *Repair or replace machines with an excessive amount of backlash.*

Machines with an excessive amount of backlash are awkward if not outright dangerous to use. Excessive backlash can allow the workpiece to suck into the cutter and possibly ruin the work.

Installing a new bronze drive nut on a worn machine will eliminate much of the backlash.

25. *Bore holes before reaming to get a straight, accurate hole.*

This applies to jobs you're doing in a lathe as well as in a mill.

Drill bits almost always walk off location a little bit. Especially when drilling deep holes with long slender bits. A reamer will follow a crooked hole.

To bring a hole back on location and straighten it out, clean up the drilled hole by boring it either with a boring bar or an undersize end mill. Then you can ream the hole if need be to get a straight, accurately located hole.

This method works well for plates and parts that need to be lined up precisely such as die shoes and mold plates.

26. *Leave small amounts of material for reaming.*

With small reamers up to about a half-inch in diameter, try to leave no more than .005″ on the ID for clean up. You can leave more but your reamed hole may go oversize. This is especially true with reamers that are a little dull. That's because as they resist cutting they try to push sideways.

With tiny reamers under an eighth inch in diameter, you should leave very little material for best results. .001″ to .002″ material on the ID for clean up should produce good results.

With reamers over a half inch in diameter, leaving .005″ to .010″ on the ID works well. A fractional drill that is one drill size under your reamer size is often convenient to use and will leave about .015″ on the ID. That is about the maximum amount of material you should leave for reaming.

Because reamers are specialty tools, you should run them slowly to avoid wearing them out. In steel, ream at about one-third the RPM you'd use to drill the hole and feed about one-third faster.

27. *Let reamers float.*

Reamers need to be able to float somewhat to produce an accurate hole. Chucking as far away from the flutes as possible within reason lets the reamer float to the center of the pilot hole and determine its own size.

FIGURE 2–4   A precision level is used to line up this lathe. Once leveled, the lathe will cut with less taper.

28. *Avoid using tape as a depth marker on drill bits.*

Using tape as a depth marker on a drill bit is risky at best. The combination of heat, chips and oil can and will push the tape farther up the drill bit which would then allow you to drill too deep.

A drill depth marker can be handy at times especially when drilling deep holes. A spot of Dykem® on a drill bit works well because it is easy to apply and holds up reasonably well.

29. *Avoid cutting with the bottom and side of your end mill at the same time.*

Do one or the other. This suggestion is especially important when using long slender end mills because of their tendency to flex under cutting pressure.

I learned this one the hard way. I was taking light cuts on the bottom and side of a mold base pocket in a conventional mill when my three-quarter inch end mill decided to grab, dig-in and take an unplanned hike up the side of the pocket. It was all over in a few milliseconds and when it was over I had broken an end mill, butchered a mold base and knocked my milling machine grotesquely out of tram. It's not funny.

30. *Glance through the general notes of your drawing before beginning a job.*

    Occasionally an engineer or draftsman will "bury" important information in the notes that you wish you had read before doing the job. One example of note that may bite you is as follows: 1) Grind to final size after heat treat.

31. *Level your lathe for a truer cutting lathe. (See Fig. 2-4)*

    Use a precision level to level the ways of a lathe. You'll be amazed at how much less taper the lathe will cut.

32. *Cut shafts precisely parallel by lining up the tailstock. (See Fig. 2-5)*

    Shaft taper is an issue machinists always seem to be fighting. One reason a lathe will cut taper on a long shaft when using a live center is because the tailstock is not centered.

    Mount an indicator in the spindle and sweep the inside of the tailstock taper. This allows you to see how far off center the tailstock is and make adjustments. Note that the tailstock would be off center by half the amount of the total indicator reading.

FIGURE 2-5   An indicator is rotated in the spindle to check the tailstock lineup.

FIGURE 2–6   An indicator is used to level the plate in the "Y" direction before cutting the ramp.

33. *Cut concentric surfaces in one setup.*

Plan your job so that concentric surfaces get cut or ground in one setup.

34. *Glue buttons on your calculator.*

I used to occasionally make mistakes doing trig problems by inadvertently pushing the button that would get me in radian mode instead of degree mode. I solved that problem once and for all by using super glue to glue the mode changing button in place on my calculator.

35. *Use an indicator to line up a thin plate before cutting a ramp. (See Fig. 2-6)*

Since you can't set a part on parallels when cutting an angle on the end of a thin plate, you must indicate the plate parallel in the "Y" direction in the vise to end up with an unskewed ramp.

36. *Machine concentric features with a lathe.*

Concentricity is easier to maintain in a lathe than a mill. Try machining a precisely concentric hole in the end of a shaft with a lathe and a mill. Odds are the lathe cut hole will be more concentric.

37. *Use digital calipers instead of dial calipers.*

Once the gears skip in your dial calipers and you get an incorrect reading, you'll never fully trust them again. At least I don't.

FIGURE 2–7   "Cosine error" is induced into indicator readings if the tip is set at steep angles relative to the surface being checked.

38. *Use the correct length tip in your horizontal indicator.*

The tip of a horizontal indicator has to be the appropriate length required or your readings will be inaccurate. It may not matter if you are indicating to find the center of hole but if you're checking the depth of a mold vent, for example with an indicator mounted with a longer tip than the indicator was designed for, your readings can be significantly wrong.

39. *Avoid cosine error. (See Fig. 2-7)*

You can avoid "cosine error" with a horizontal indicator by keeping the tip of the indicator extended more or less parallel to the surface you are checking. If you take readings with the tip of the indicator positioned at steep angles as shown in the photo, the accuracy of your readings will suffer. Readings will be 100% off when the tip of the indicator is set 60° from horizontal. In other words, if you compare parts that are .001″ different in height with the tip set at 60°, the dial reading will show .002″.

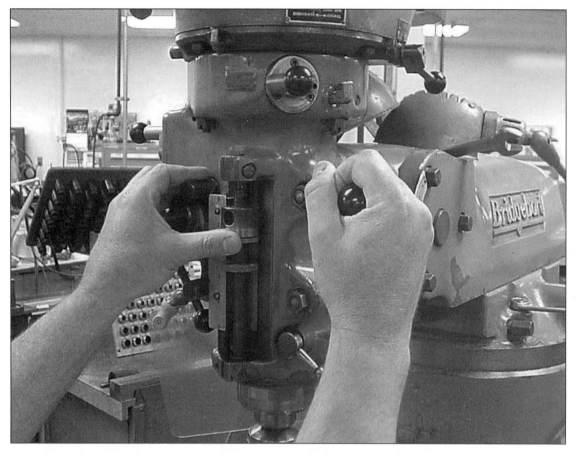

FIGURE 2–8   Pecking against the quill stop is the best way to feed small cutters in "Z".

40. *Avoid chamfer errors on lathe parts.*

Assuming one mark on your dial will cut a thousandth off a diameter, how far do you have to plunge in the "X" direction which is perpendicular to the ways of the lathe with a 45° form tool to get a .030″ X 45° chamfer? If you said .030″ you would be wrong. The answer is .060″. How far do you have to plunge in the "Z" direction which is parallel with the ways of the lathe with a 45° cutter to get a .030″ X 45° chamfer? The answer is .030″.

The thing you have to remember is that the cross slide on a lathe actually moves half the amount that would be taken off a diameter.

41. *Feed small drills and cutters against the quill stop. (See Fig. 2-8)*

Use the quill stop to feed against when machining with tiny drills and cutters. Rotate the quill stop as you peck against the work plunging about .001″ at a time. If you feed tiny drills and cutters by hand without using the quill stop, you'll likely end up breaking them. This technique also works well for starting cutters into surfaces that are not perpendicular to the cutter.

42. *Keep track of the 0, 0, corner of blocks.*

When you have multiple square or rectangular blocks to machine, it is best in terms of consistency and accuracy to keep track of the 0, 0, corner of the blocks so that all features are machined relative to that corner. One way to mark the 0, 0, corner of blocks is by applying a small spot of Dykem®.

43. *Give your machine's digital readout a once over.*

Digital readouts are great tools when they work.

To make sure your digital readout isn't skipping, compare the readings from your digital readout to the readings on your dial. Even though a lead screw and drive nut wear, the thread pitch of those items remains constant so that handle dial readings remain relatively accurate in spite of the wear.

Start by zeroing both the digital readout and a handle dial. Then crank the machine a few inches to see if your readout and dial readings stay close together. Continue the process until you are satisfied that the readout is not skipping.

If you find that the readout is skipping, let everyone in the shop know so that nobody ends up making junk.

44. *Avoid using too much tolerance.*

One common mistake beginners make is using too much tolerance on seemingly unimportant features. Often machinists need to hold parts for subsequent machining operations using features or surfaces that have already been machined. If those surfaces have been machined erratically to varying dimensions perhaps within some wide open tolerance then it will be difficult to use those surfaces later to accurately locate the parts for further machining.

45. *Avoid clamping on a tool radius. (See Fig. 2-9)*

Many cutting tools such as end mills and fly cutters have shanks that blend into the body of the tool with a fillet radius. In terms of strength and rigidity, that makes sense but the radius can get you in trouble if you're not careful. If you inadvertently stick the tool too far into a holder, you may end up clamping on the radius which can cause the tool to come loose when it is cutting. Clamp tools at least 1/16″ away from shoulder radii.

46. *Keep your indicator tuned up.*

Make sure your indicator point is tightly screwed in and doesn't have any lateral play.

FIGURE 2–9   Clamping on cutter shank radii can cause cutters to come loose. It's a condition that must be avoided.

47. *Set facing tools precisely on center.*

There's no way around it. Your lathe facing tool has to be set to cut precisely on center. If not, there is a high probability the tip of the tool will break as it approaches the center. Either that or it will leave a tit on the end of the part. I've always used the trial and error method for setting facing tools by taking light cuts and adjusting the height of the tool accordingly.

48. *Maintain close nominal sizes on parts held in collets.*

When using collets, make sure you maintain a close tolerance on diameters that may be used to hold or locate the parts later on. Also if possible, turn your parts to a nominal collet size. For example, if your print calls for a feature to be .44″ in diameter, then you are better off from the standpoint of fitting a collet to make the diameter .4375″.

The common 5C collets are sensitive to variations in part diameter for a couple of reasons. The first is because a part whose diameter does not fit the collet properly will tilt in the collet.

FIGURE 2–10 The ram of a mill must be indicated parallel to the "Y" axis when the head is tilted to avoid machining skewed holes and features.

Secondly, parts that vary in diameter even by a few thousandths "suck in" different distances along the "Z" axis when the collet is closed. Any variances in diameters will cause a corresponding but magnified variance in shoulder distances.

49. *Double check gauge block stack-ups with calipers.*

A common use of gauge blocks is to set sine plate angles. The stacking of gauge blocks is nothing more than a series of calculations, which like any calculation, is prone to error. It is a good idea to double check overall stack-ups with calipers to make sure you have the stack-up you want.

50. *Indicate the ram of a milling machine parallel to the "Y" axis before machining with the head tilted. (See Fig. 2-10)*

The ram must be indicated parallel to the "Y" axis before machining with the head tilted; otherwise holes and features will end up skewed in the work.

FIGURE 2–11 Parts are wrapped before heat treating to prevent oxidation.

51. *When cooling heat treated parts avoid setting them directly on the ground.*

    It is best to set them on an elevated screen that allows for consistent cooling all around.

52. *Wrap parts in foil for heat treating. (See Fig. 2-11)*

    To prevent oxidation and decarburization when heat treating parts it is best to carefully seal them in foil. Type 321 stainless steel foil .002″ - .003″ thick works well.

53. *Admit your mistakes.*

    If nobody notices or cares about some mistake you've made then just forget it. If somebody shows an interest in your mess up (usually the case), then simply admit it. In most cases it will soon be forgotten. Let the mistake be a learning experience. Also, keep your screw-ups in perspective. How does your screw-up compare to the Titanic hitting an iceberg and sinking? Keep that in mind next time you make a bad cut.

chapter **3** **Do It the Easy Way**

I remember the first time I tried to line up a large plate in a milling machine. I must have been back and forth over that plate twenty times before I got it straight. I remember thinking to myself "there must be an easier way." There was, of course, I just wasn't aware of it.

In our never ending quest to find easier ways to do things, we can't help but learn a thing or two. Sometimes we'll figure things out for ourselves and other times we have to be shown. Either way, finding an easier way to do something is often rewarding. "Easier" is almost always faster and is often more accurate.

FIGURE 3–1   This indicator holder has been modified to allow the indicator to be mounted directly in its frame.

1. *Line things up the easy way.*

    If you do the procedure properly, you can get a vise or a plate lined up with one or two passes of your indicator.

    Start by lightly tightening one side of the vise to the mill table. Run an indicator slowly from the snug side of the vise towards the loose side, tapping the vise in as you go. If you get to the end of the vise jaw before it is square then go back (rapid traverse) to the snug side and repeat the process. Don't try to tap the vise in by going from the loose side the snug side. You'll probably get messed up. After the vise is square then tighten both sides.

2. *Tram the head of a conventional mill with a hammer.*

    This is an operation that takes a little practice. One thing that makes tramming difficult is that the head will almost always move a little when you tighten the nuts that hold the head to the ram. There is a way to avoid that problem. Once the head is close to being trammed in, lightly tighten the nuts then use a soft hammer to tap the head in precisely. By using this method, the head won't move when you put the final torque on the nuts. Before tapping the head in, release any pressure you have on the head rotation gears.

FIGURE 3–2   A grinding vise can be mounted in a milling machine vise to quickly rotate parts ninety degrees.

3. *Put a thumbscrew in the frame of your Indicol® to make it more versatile. (See Fig. 3-1)*

   Indicol holders are handy when you want to indicate something without removing a cutter from the spindle. When the long gooseneck doesn't fold up sufficiently to position the indicator correctly, you have to find some other way to hold the indicator.

   A hole in the frame of the Indicol with a thumbscrew to hold the indicator in place increases the versatility of the tool. You'll be able to indicate small holes and other features that are difficult to reach with the gooseneck arrangement.

4. *Indicate large diameters when you have the option.*

   When setting up round parts in a mill or lathe with an indicator, it is easier due to the gradual change in indicator readings, but not necessarily more accurate, to sweep-in larger diameters.

5. *Auto feed parts to length in a mill using a grinding vise. (See Fig. 3-2)*

   Let's face it, we'd all rather auto feed than hand crank when given a choice. If your milling machine doesn't have an auto feed in the "Y" direction you can still use your "X" axis auto feed to machine the ends of bars. One way is to use a grinding vise

FIGURE 3–3   An angle is being cut on this part by rotating the spindle in reverse and using the lathe compound.

FIGURE 3–4   Emergency collets are the best way to hold thin washer type parts.

clamped in the regular milling machine vise which rotates the part ninety degrees so that you can use the "X" axis auto feed.

Some old timers may think using a precision grinding vise that way is primitive. I see nothing wrong with it.

FIGURE 3–5   A screw mounted in the end of a collet stop allows the stop to be used in smaller diameter collets.

6. *Run a conventional lathe in reverse when cutting angles with the compound. (See Fig. 3-3)*

   To position the crank of the compound toward the operator side, run the spindle in reverse and cut the angle on the far side of the part.

7. *Use emergency collets to make thin, washer type parts. (See Fig. 3-4)*

   Emergency collets are soft so you can bore them to whatever size and depth you need. Modifying emergency collets is the easiest and most accurate way to make thin, washer type parts. You simply bore and face the collet to a depth slightly less than the thickness of the part and away you go.

   I've found that the cheap imported emergency collets work fine. Be sure to use a boring bar with a tiny tip radius so that parts lie flat against the back of the counter bore without fillet interference.

8. *Put a small screw in the end of your collet stop to make it more versatile. (See Fig. 3-5)*

   A small screw mounted in the end of your collet stop allows you to use the stop in small diameter collets.

9. *Use ultra thin parallels to avoid drill interference.*

   Many parts have holes in them that come close to the edges of the part. You can purchase 1/32″ thick parallels to eliminate most drill bit interference problems.

FIGURE 3–6   A small V-block is used to nest a boring bar in this lathe tool holder.

10. *Use a small V-block to hold a small round boring bar in a lathe tool holder. (See Fig. 3-6)*

This is a fast, simple, sturdy way to hold small round boring bars in lathe tool holders. I use this arrangement often.

11. *Rough cut slots by plunging in "Z". (See Fig. 3-7)*

Slots are commonly used in machined parts. One way to make an accurate slot with a conventional mill is to first drill then bore each end of the slot to finished size with an on-size end mill. Then go back with an undersize end mill and rough out the center section by plunging the end mill in "Z" as you step over. This plunging method works well for quickly removing material. An end mill has a tendency to pull to one side when plunging this way so you may want to bias the end mill to one side. Once the center section has been roughed out, you can go back and finish the slot with an on-size end mill.

FIGURE 3–7   Slot milling is made easier by boring on-size holes at each end of the slot before roughing out the center section.

12. *Use a screw plate to cut and file small screws to length. (See Fig. 3-8)*

The beauty of using a screw plate to file small screws to length is that when the screws are removed from the plate the threads are automatically cleaned up. This is a tool you'll probably have to make. I am not aware of any manufacturers that make these plates. A hardened screw plate will last longer.

13. *Use a fly cutter to cut holes in sheet metal. (See Fig. 3-9)*

Sharpen a cutter like a grooving tool and set it at the proper angle in a fly cutter to cut clean holes through sheet metal.

FIGURE 3–8   A screw plate is used to file a small screw to length. When the screw is removed, the threads are automatically cleaned up.

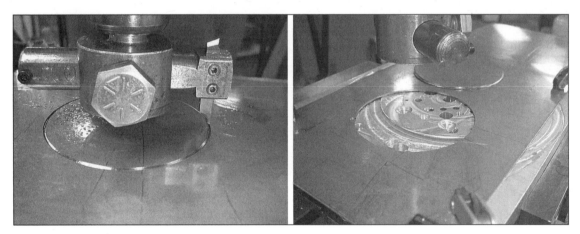

FIGURE 3–9   A fly cutter is used to cut a hole in sheet metal.

14.   *Use double sticky tape to hold down thin sheet parts for machining. (See Fig.3-10)*

Double sticky tape can be used to hold down parts if the part has a lot of surface area. Double sticky tape works well when the pressure of the cut forces the material down onto the tape such as when fly cutting. In some cases you may be able to do a little side milling such as pocketing if you keep cutting pressures light. You can not use coolant with double sticky tape since the water will cause the tape to release. When

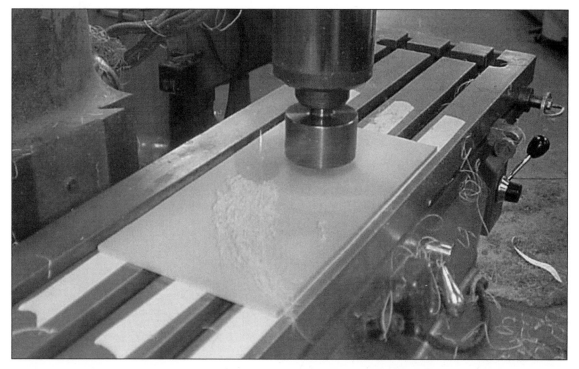

FIGURE 3–10 Double sticky tape is an effective way to hold thin sheet stock for fly cutting.

FIGURE 3–11 Perishable subplates are handy for holding parts and machining them beyond the bottom of the part for complete cleanup.

using end mills for pocketing it is best to avoid letting the end mill cut into the tape so the end mill doesn't get gummed up.

15. *Use subplates to allow for over cut. (See Fig. 3-11)*

I'm a believer in using subplates. They allow cutters and drills to cut through parts for 100% cleanup. They also allow for easy part mounting. They can be used numerous times on different jobs and can be resurfaced when necessary.

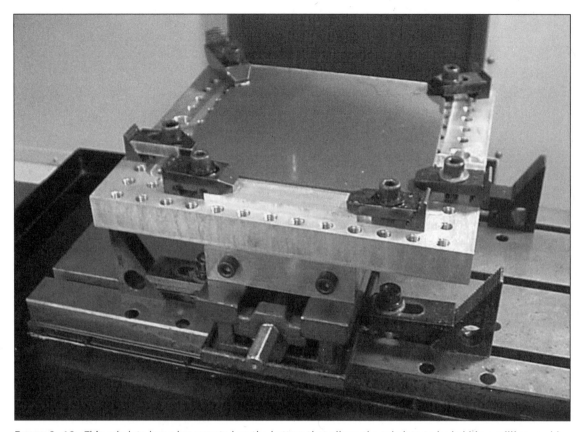

FIGURE 3–12 This subplate has a bar mounted on the bottom that allows the subplate to be held in a milling machine vise.

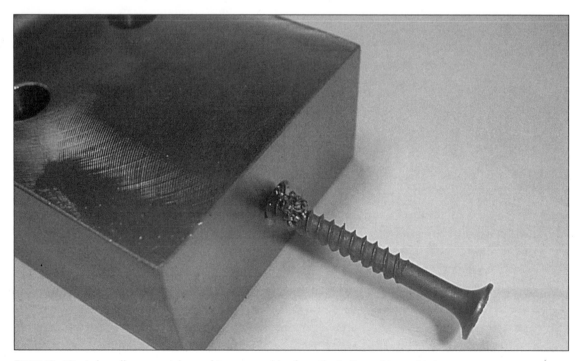

FIGURE 3–13 A drywall screw can be used to remove chips from the bottom of tapped holes.

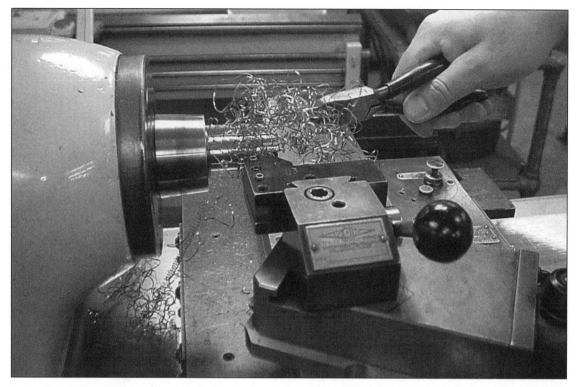

FIGURE 3–14 The ability of wire cutters to sever chips makes them useful for removing long, stringy lathe chips.

16. *Make a subplate that can be mounted in a vise. (See Fig. 3-12)*

   A large subplate with a hefty rail mounted on the bottom allows for quick use of the plate in a milling machine vise.

17. *Use preexisting angles instead of a sine bar for setting angles.*

   Angles come in sets ranging from one to forty-five degrees. You can stack the angles to get different degree combinations. Use a sine bar only when you have to.

18. *Use a drywall screw to remove chips from the bottom of tapped holes. (See Fig. 3-13)*

   Compressed air may not always remove chips from the bottom of tapped holes. Drywall screws grab chips very well and make removing them relatively easy.

19. *Use diagonals or wire cutters to pull chips away from lathe work. (See Fig. 3-14)*

   I like to avoid producing long stringy chips when I can. In fact, I go out of my way to find the right combination of tool geometry, speed, feed and depth of cut that won't give me long stringy chips, but I don't always succeed. When I don't succeed, I use diagonals or wire cutters to pull and cut chips away from the tool and workpiece. The ability of diagonals to sever chips makes removing them easier.

FIGURE 3–15 Modified dowel pins are mounted in the table slots to create a stop parallel to the "X" axis.

FIGURE 3–16 Two vises mounted on a machine table are handy for machining long parts.

20. *Use modified dowel pins to create a "stop" parallel to the X-axis on a milling machine table. (See Fig. 3-15)*

Most small milling machine tables have 5/8″ wide T-nut slots. An easy way to create a stop parallel to the X-axis is to install a pair of 5/8″ dowel pins in the slots. That works well except for one issue. They may be difficult to get out. If you grind or cut flats on the sides of the dowel pins then you can rotate the dowel pins in and out of the slots with an adjustable wrench. I've seen people express minor amazement after they see how easy it is to remove these modified dowel pins from the T-nut slots.

21. *Mount two vises on a milling machine table. (See Fig.3-16)*

This is an effective way to hold and machine long parts. When setting up two vises on a machine table you may need to shim one of the vises to make their heights match.

22. *Use long parallels to support long parts. (See Fig. 3-17)*

23. *Rotate parts in a vise to dress edges. (See Fig. 3-18)*

If you are dressing the edges of a part using a chamfer or radius tool in a conventional mill, the easiest way to dress all the edges alike is to use the back jaw of the vise as a reference plane. Set your cutter correctly one time relative to the part and back jaw then rotate the part to dress the edges. All of your dressings will come out the same that way with minimal effort.

FIGURE 3–17 Long parallels can be use to support long parts.

FIGURE 3–18 The easiest way to dress edges in a conventional milling machine is to rotate the part in the vise.

FIGURE 3-19 A small drill bit is used to line up a prick punch mark.

24. *Use a small drill to pick up a punch mark. (See Fig. 3-19)*

One accurate way to pick up a punch mark is to do it in a milling machine using a small drill bit. (.030″ to .040″ dia.). When gently brought into contact with a punch mark, a tiny spinning drill bit will flex as it tries to follow the crater left by the punch. By adjusting the table until the drill bit stops flexing you can be sure the spindle is over the center of the punch mark.

25. *Hand drill a perpendicular hole. (See fig 3-20)*

Once in awhile you will be faced with a situation where you need to accurately hand drill a hole. One of the easiest ways to do that without the aid of fixturing is to have another person help you. For example, if you need to drill a hole accurately in the side of a mold base, using this technique you would have someone stand to the side of you and eyeball the up and down position of the drill while you as the driller eyeball the side to side position. It is best to start with a drill bit that is somewhat undersize. As simple as this method sounds, it works great.

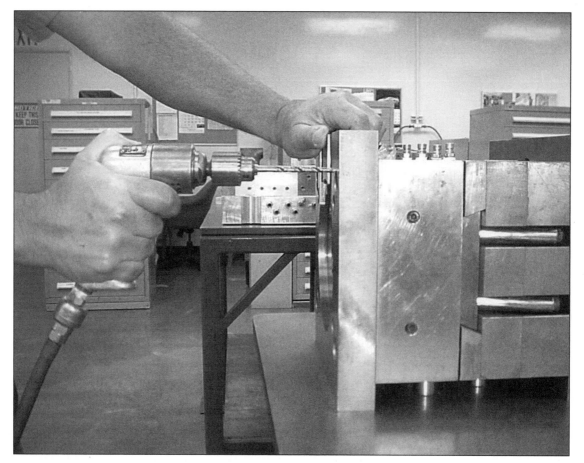

FIGURE 3-20 A perpendicular hole is hand drilled with the help of a spotter.

26. *Use plug gauges or gauge pins to accurately measure inside diameters. (See Fig. 3-21)*

    It's hard to beat a plug gauge or gauge pin to precisely measure the ID of a hole.

    Telescoping gauges take skill to use and are prone to error. I prefer using or turning a plug gauge for measuring bore size when extreme accuracy is needed.

27. *Change sanding disks using wax paper as shield against premature sticking.*

28. *Use sharp tools that cut with light pressure to hold tight tolerances.*

29. *Avoid using talcum powder on surface plates.*

    Talcum powder gums things up quickly. The best way to keep a surface gauge sliding freely is to keep the gauge and the surface plate clean. Glass cleaner works fine.

FIGURE 3–21 It's hard to beat a plug gauge for precisely measuring the ID of a hole.

FIGURE 3–22 A band saw blade is silver soldered to create a strong, durable joint.

30. *Silver solder band saw blades. (See Fig. 3-22)*

Most band saws come with a blade welder. If you've been working in a shop for any length of time you've probably seen somebody or been in the position yourself of carefully welding, grinding and annealing a blade only to have it break within seconds of installing it back in the saw. These "homegrown" welds are inconsistent at best.

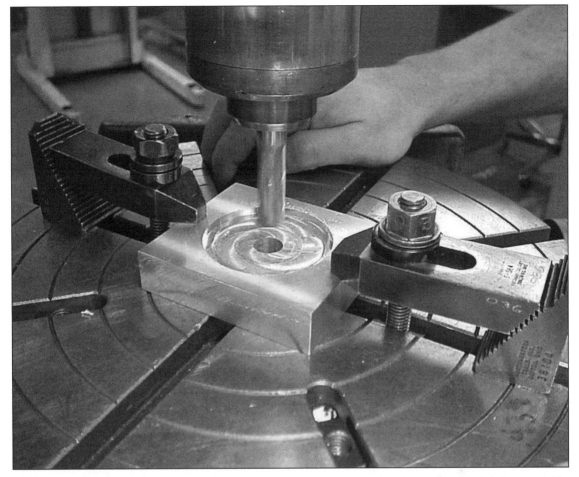

FIGURE 3-23 A large counter bore can be machined with a rotary table. In this picture the table is being rotated by hand.

I've all but given up on band saw blade welders. In my experience silver soldering a blade together is stronger and more consistent.

If you bevel both ends of the blade at a about a five degree angle and silver solder them together, you'll have a joint that will likely hold for the life of the blade.

As a side note, I believe many people over tighten and put undue stress on band saw blades. The blade only has to be tight enough so that it doesn't wobble or run off when cutting.

31. *Use the rotary table to cut large counter bores. (See Fig. 3-23)*

A large counter bore is one of the more difficult features to quickly machine with a conventional machine. There is an expensive boring head on the market that is capable of boring and facing the bottom of a counter bore however they're a bit of a hassle to use. One of the easiest ways to machine a large counter bore with a conventional machine is to do it in a rotary table.

32. *Avoid using reamers to size holes in bearing bronzes.*

Bearing bronzes are difficult to ream because the materials have a tendency to squeeze down on reamers. It is easier to use a boring bar to bore holes to size in these materials.

33. *Machine holes in milled parts right after they've been squared.*

The best time to machine holes in parts is right after they've been squared because that's when the parts are easiest to

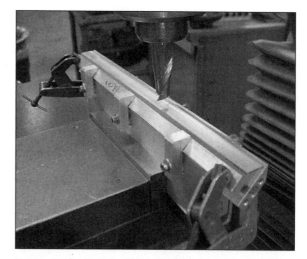

FIGURE 3–24 Well designed fixtures take the aggravation out of running jobs.

hold and edge find. Once a bunch of features have been milled into parts, they usually become more difficult to work with.

34. *Use Kant-Twist® clamps. (See Fig. 3-24)*

What a great invention. They've almost made C-clamps obsolete.

35. *Make fixtures for easier machining. (See Fig. 3-24)*

Often during the planning stages of a job, one is faced with the dilemma of whether one should take the time to make a fixture to facilitate the machining or just struggle through the job with an awkward setup. The choice usually boils down to time. Once I determine there is no advantage time wise to either choice then I'll almost always go with making a fixture. I dislike fighting lousy setups.

36. *Deburr parts in a tumbler. (See Fig. 3-25)*

Tumblers don't work very well for removing heavy burrs but they work well for rounding sharp corners. If you knock down large burrs with a file before putting parts in a tumbler they will come out with slightly radiused corners all around that need no additional filing. I've had good results leaving parts in overnight using medium grit stones.

37. *Drill deep holes by using progressively smaller drill bits.*

This method works great for drilling deep flushing holes in graphite electrodes and other materials that have a tendency to squeeze down on cutters.

FIGURE 3–25 Tumblers don't work very well for removing heavy burrs but they work great for rounding sharp corners and edges.

To relieve friction and binding, use progressively smaller drill bits as you drill. Once you are to depth with a smaller drill, it is easy to go back and finish the hole with the on-size bit.

38. *Use a shim to make plates come out parallel.*

If you are fly cutting a plate in a milling machine and the plate doesn't come out parallel in the "Y" direction; you can quickly correct the error by placing a shim under the thick side of the plate. If you use a shim the thickness of the error then re-cut the plate, it should come out parallel.

39. *Cut an angle on the edge of a plate without tilting the head of your milling machine. (See Fig. 3-26)*

With these setups you can quickly cut angles on the edges of small plates and bars. It is best to avoid tilting the head of a mill whenever possible so that you don't have to re-align it.

FIGURE 3–26 Two setups are shown that can be used to make angle cuts. You should avoid tilting the head of a mill when possible.

FIGURE 3–27 The tapped holes in the edge of this angle fixture provide a variety of stop positions for locating various size parts for angle machining.

40. *Construct an angle with tapped holes in the edge to provide a variety of stop positions for locating various size parts. (See Fig. 3-27)*

An angle with tapped holes in the edge can be used to support and locate various size parts for machining.

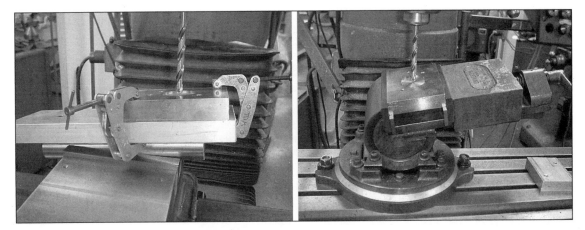

FIGURE 3–28 A simple universal angle fixture such as the one shown on the left can be used to hold parts for drilling and machining compound angles. A universal angle vise such as the one shown on the right can also be used.

41. *Construct a poor man's universal angle fixture. (See Fig. 3-28)*

Occasionally you may need to drill a hole or machine something on a compound angle. There are a variety of ways to do compound angle machining such as tilting the head of a mill, mounting the work on a compound sine plate, or mounting the work in a universal angle vise such as the one shown in the right photo. One easy to hold something on a compound angle is to use a poor man's universal angle fixture such as the one shown in the left photo. The fixture is nothing more than a plate mounted on top of a round bar. The fixture can be rotated and tilted in a milling machine vise to provide a wide range of angles.

To drill a hole on an angle you must first machine a flat seat into the work using an end mill which is at least as large as the diameter of the hole you want to drill. Then you can center drill and drill as you would normally.

42. *Drill a pilot hole using a drill bit about one third final diameter.*

Too much pressure applied to a drill bit can cause bad things to happen. A pilot hole significantly reduces the pressure needed to drill a larger hole.

43. *Use a piece of paper to position a rotating end mill. (See Fig. 3-29)*

A piece of copy paper is about three thousandths thick. You can find the location of a surface without gouging it by using a strip of paper. Hold the paper between the end mill and the workpiece while approaching the cutter to the workpiece. The paper will start to pull and shred when the end mill is within a few thousandths of the workpiece.

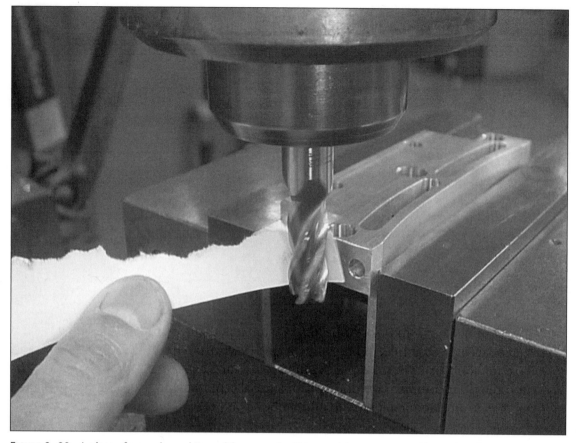

FIGURE 3–29 A piece of paper is used to position an end mill tangent to a part to avoid gouging the surface.

44. *Use an indicator instead of an edge finder to find the exact center of a shaft. (See Fig.3-30)*

Sweeping an indicator over either side of a horizontal shaft is the most accurate way to find the center of the shaft.

45. *In a milling machine, use an indicator and calculator to find the exact center of a plate. (See Fig. 3-31)*

This is a very precise way to find the exact center of a plate and is the best method to use when extreme accuracy is needed. A digital readout in good working order is essential for accurate results.

Start by mounting a horizontal indicator in the spindle and positioning it off to one side of the plate. Sweep or rotate the indicator point across the side of the plate so that you get a few thousandths preload on the point. Set the indicator dial to zero at the point of greatest deflection then zero the digital readout. Avoid touching the indicator for the remainder of the process. Raise the indicator then move the table so the spindle goes to the opposite side of the plate. Sweep the indicator across this opposite side while carefully moving the table until the indicator reads zero again. Make a note of the total table travel shown on the digital readout. That number divided by two is the

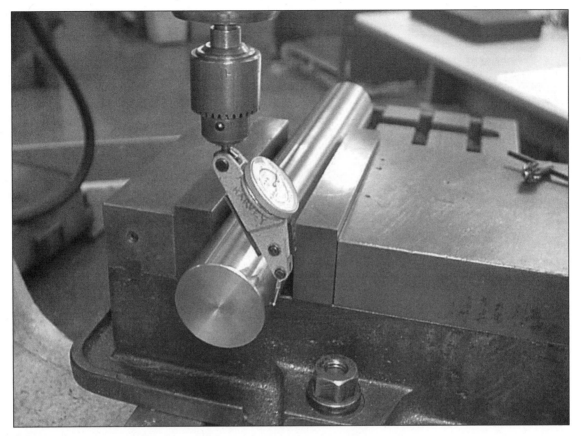

FIGURE 3–30 An indicator is used to precisely find the center of a shaft.

FIGURE 3–31 An indicator is used to precisely find the center of a rectangular part.

exact center of the plate in that axis. Repeat the process for the other axis to locate the exact center of the plate in both axes.

This same principal can be used with an edge finder to find the center of a plate and for beeping off an electrode in an EDM machine to find the center of a plate.

FIGURE 3–32 Preexisting angles can be used to set vise angles.

46. *Indicate a milling machine vise at an angle using pre-existing angles. (See Fig. 3-32)*

47. *Make a set of flat bottom drills. (See Fig. 3-33)*

    One of the easiest ways to remove the angle left at the bottom of a hole by a drill bit is to use a flat bottom bit of the same diameter. Flat bottom bits can be made from regular drill bits by grinding the tips square then adding a little relief to the flutes.

48. *Reverse bore a boss in a milling machine to avoid making a time consuming setup in a lathe. (See Fig. 3-34)*

49. *Before spending time to precisely line up and clamp a large part on a machine table, make sure the table has enough travel to do the job.*

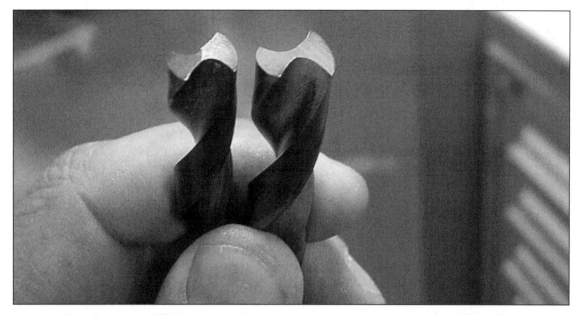

FIGURE 3–33 A flat bottom drill bit can be used to remove the lead angle left by a standard drill bit. The flat bottom bits in this photo were made from standard bits ground flat and relieved.

FIGURE 3–34 Sometimes you can avoid time consuming setups in a lathe by reverse boring parts in a milling machine.

50. *Purchase large sheets of material in 4' X 4' sections instead of 4' X 8' sections for easier handling.*

51. *Purchase a geared head drill press for rapid spindle speed changes.*

    Geared head drill presses beat the heck out of belt driven drill presses when it comes to quickly changing spindle speeds.

52. *Cut your own raw stock to length.*

    I used to work in shop where someone other than the machinist cut and prepared raw stock for machining. I never liked that policy simply because it limited the way a machinist could run the job.

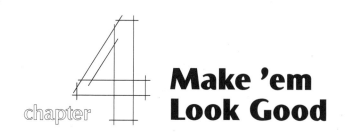

Make 'em Look Good

Why do we want our parts to look good? One answer is that we usually have to sell them, either to an inspector or customer. Parts that have a skillfully made appearance are nearly always more accurate, more consistent and easier to sell than carelessly made parts.

Another reason clean, detailed parts are desirable is because they are easier to measure, inspect and talk about. If someone were to ask "How deep is the slot in that part?" it is easy to say "It's within a couple thousandths of a half inch deep." It's not so easy to explain that for a variety of reasons, you're unable to get consistent readings.

People may argue that appearance shouldn't matter as long as parts are in tolerance. Ragged parts can be difficult to inspect to determine if they are in tolerance.

What constitutes quality part appearance? My concept has evolved over the years.

The following are a list of criteria I use to judge part appearance before I ever put a micrometer or caliper on them.

- Does the finish correspond to the function of the part?

- Are finishes consistent throughout the part?

- Is the part deburred properly?

    Minimum and consistent deburring is functional and looks better than heavily filed edges.

- Is the part covered with chatter marks?

    A little chatter may not be an issue on milled parts and can often be seen in long corners and weak areas. Try to keep chatter to a minimum though or it'll look like you don't know or don't care what you are doing. Chatter can almost always be reduced or eliminated by reducing spindle speed.

- Are "blended" cuts mismatched? (See Fig. 4-1)

    Occasionally a surface has to be machined with the cutter coming onto the surface of the part from different directions, usually to avoid a feature of some sort. Sometimes it can be difficult to blend cuts into a flat, continuous surface. If blending is done well, you can be reasonably sure that some care was taken when the parts were made.

- Does the part sit flat on a surface plate or does it rock back and forth like an old hubcap?

    Warped parts are difficult to work with. They are an indication that the machinist either paid little attention to material strain or used a worn out machine.

- Are holes tapped sufficiently deep?

    If not, shallow threads may cause assembly problems later on.

- Are chips left in the bottom of tapped holes?

    In some cases it doesn't matter. When it does, use a dry wall screw or compressed air to remove them.

- Are surfaces smooth or do they have redeposited chips, metallic fuzz and other questionable residue that may fall or rub off later on?

- Will the customer, inspector or another person besides the original machinist have to detail or rework the parts to make them useful?

    If so, then the original machinist didn't finish the job.

    Some people argue that it takes too much time to detail parts and that it's not worth the extra effort. All things being equal they probably do take a little longer. However, that extra time is usually more than offset by the ease with which the parts are setup, machined, inspected, sold, used and if necessary reworked.

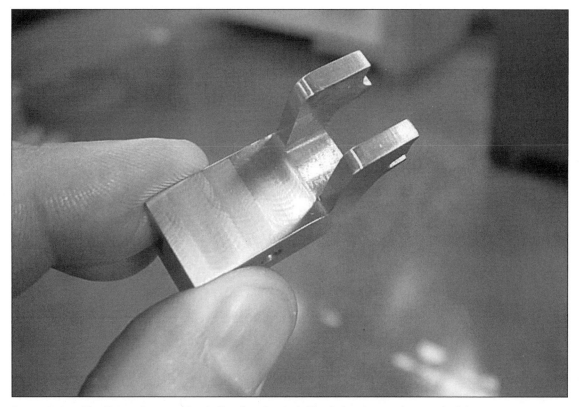

FIGURE 4–1   Blending surfaces can be challenging. Precisely blended surfaces make parts look better.

The following suggestions will help you produce quality parts without sacrificing much time.

1. *Use a fly cutter to produce smooth finishes.*

   I've heard it said that "anybody that uses a fly cutter is not a real machinist." Aside from being a vague and simplistic statement, I tend to disagree with it. There are advantages to using a fly cutter. First, they last forever. When the cutter gets dull it is a simple matter of hand sharpening one edge or tip. Second, they can be adjusted to match the width of the part. Third, they leave a good finish.

   The disadvantages to using a fly cutter are that they are generally less rigid than other cutters and to get a nice finish you have to feed relatively slowly.

   Bear in mind that since they are normally used as finishing tools, the fact that they have to be fed slower is not a great disadvantage. Most tools needs to be fed slower on finishing cuts.

   When cutting steel with a fly cutter, I prefer using a cutting tool with a relatively small tip radius. A 1/32″ radius works well on most steels. A 1/32″ radius cuts with fairly low tool pressure yet leaves a smooth finish. If the radius is much smaller the cutter may leave pronounced tool feed marks on the part. If the radius is much larger, the

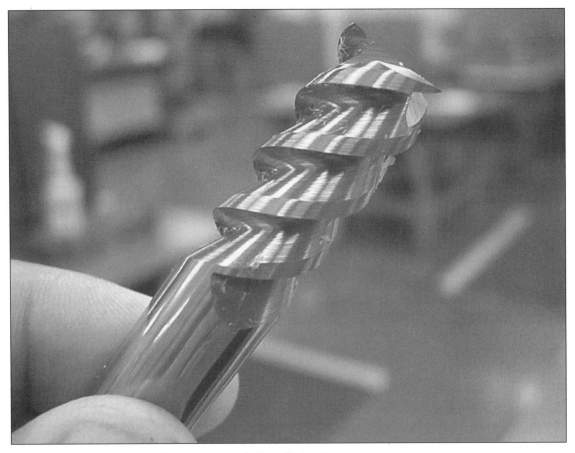

FIGURE 4–2   High helix end mills produce smooth side milled surfaces.

increased cutting pressure caused by the large contact area may cause excessive tool deflection and chatter.

Since aluminum is softer, you can get away with using a larger radius which allows you to feed faster without leaving pronounced tool marks.

2. *Avoid side milling when you can.*

A fly cut or face cut surface will almost always look better and be flatter than a side cut surface. Side milled surfaces are prone to more ailments than face cut surfaces. The finish of a side-milled surface is based on, among other things, the condition of the cutting edges over the length of the end mill. If the end mill has some nicks in it then it may leave tracks in the part. Also, a long end mill almost always flexes away towards the bottom of thick parts; more so when the cutter is dull. Another factor is end mill taper which is more common with reground end mills. Whatever taper an end mill has will show up in the part.

Sometimes the additive effects of flexing away, end mill nicks and end mill taper will produce quite an inaccurate, lousy surface.

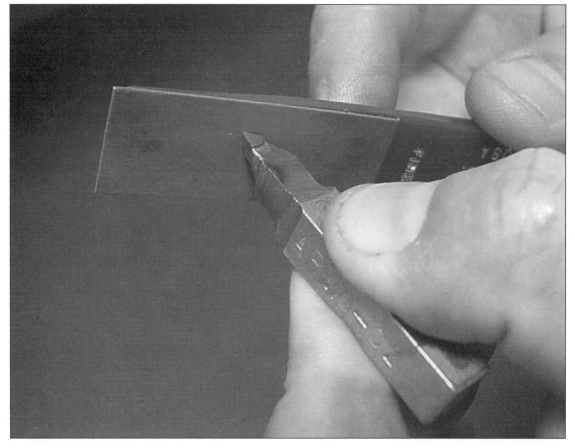

FIGURE 4–3   A diamond lap is used to hone a smooth round cutting tip on this lathe tool. This lap was made by EZE-LAP Diamond Products.

When milling pockets you can't avoid side milling. In that case, it is best to use a new or sharp end mill to finish the pocket. For external finishing cuts, fly cutting or face cutting will almost always give you a better finish.

3. *Use sturdy high helix multiflute end mills when side milling to achieve the best surface finishes. (See Fig. 4-2)*

High helix multiflute end mills produce superior surface finishes to standard end mills. Use them when smoothness and accuracy are important.

4. *Use a flat diamond lap to hone cutting tools. (See Fig. 4-3)*

Smooth round cutting points last longer and provide better finishes than points with sharp corners or facets.

It is difficult to grind a smooth round cutting point by hand; especially the small points like the .015″ tip radii often used on finishing lathe tools. You can lap a smooth cutting tip using a flat diamond lap. Diamond laps work well on just about any type of

FIGURE 4–4   Scotch-Brite® has multiple uses in a machine shop. In this photo it is used to smooth rough edges.

hard steel including high-speed steel, cobalt and carbide. They can be used on end mills as well as lathe tools.

When applying a tip radius to a lathe tool, lightly break the cutting point with a grinder to start the tip radius. Use a diamond lap to hone the point to a smooth radius. Move the diamond lap laterally around the point of the tool to get a smooth radius and maintain a straight, consistent relief angle.

A diamond lap cuts better and leaves a sharper edge than a stone.

5. *Use Scotch-Brite® to remove residue and metallic fuzz from the surface of a part. (See Fig. 4-4)*

It would be nice if we could always end up with perfectly smooth surfaces on final cuts. Sometimes it doesn't happen. One way to quickly smooth out and clean up fuzzy surfaces is by going over the surface with a piece of Scotch-Brite®. This method is especially useful for lathe parts.

FIGURE 4–5   Six jaw chucks spread clamping pressure over a greater area than three jaw chucks. Six jaw chucks are great tools for holding lathe work.

6. *Use the smallest diameter end mill you can get away with to side mill parts.*

    By using small diameter, multi-flute end mills, you'll get the best possible finish. Relatively speaking, the smaller the end mill the faster it can be rotated before wearing out. Faster spindle speeds give better surface finishes at a given feed rate. The end mill you choose must be large enough and stiff enough so that it doesn't flex when cutting.

    You can use the tan chip rule as a baseline for determining feeds and speeds when side milling. Remember, the tan chip rule applies to cutting ferrous alloys in a dry condition.

7. *Use a six-jaw chuck in a lathe for gripping finished surfaces. (See Fig.4-5)*

    Sometimes you are forced to grip on a finished surface to machine other features of a part. A six-jaw chuck spreads clamping pressure over a larger area than a three-jaw chuck and reduces the possibility of damaging the surface. Also, six jaw chucks don't distort thin walled parts as readily as three jaw chucks.

FIGURE 4–6   Boring soft jaws may be the only way to get parts to run concentric in worn-out three jaw chucks.

You will sometimes see machinists use small aluminum or brass pads between the jaws of a three or four jaw chuck and the workpiece to protect a finished surface. I'm not a big fan of that method and avoid it when I can since it requires a lot of fiddling around. I would rather use a six jaw chuck and take light cuts if need be to avoid putting a lot of pressure on finished surfaces.

8. *Use aluminum soft jaws to hold finished surfaces. (See Fig. 4-6)*

Another way to hold finished surfaces without damaging them is by turning a set of aluminum soft jaws to hold the part. If the part is thin walled then soft jaws may be the easiest way to go. One of the biggest problems with soft jaws is that it is usually difficult to find a set that hasn't been all cut up.

9. *Use soft jaws in a milling machine vise. (See Fig. 4-7)*

Modified soft jaws mounted in a vise can be useful for holding and nesting parts.

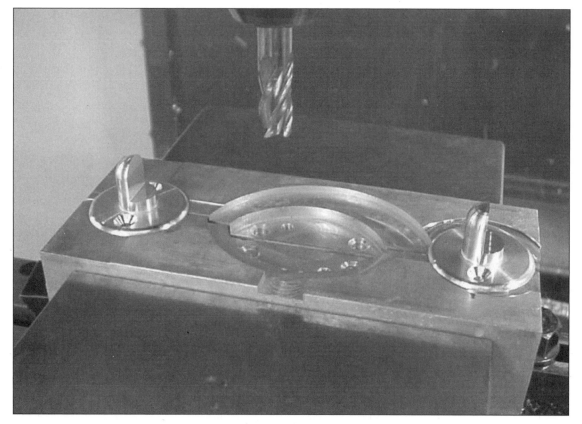

FIGURE 4–7   Perishable vise jaws come in handy for holding odd shaped parts.

10. *Prepare larger pieces of raw material with an orbital sander. (See Fig. 4-8)*

Sheet stock usually arrives from the vendor with a variety of nicks, scratches and other "undesirables." Orbital sanding is a great way to prepare sheet stock for machining. 120 or 180 grit sandpaper cleans a surface well and leaves a smooth, textured finish.

11. *Clean up machined surfaces over a sheet of sandpaper on a surface plate. (See Fig. 4-9)*

This is a great way to "tune up" machined surfaces. After parts have been handled, clamped and machined a few times some surfaces may have unwanted nicks and abrasions. Lightly sanding the surface of a part on a sheet of 500 or 600 grit sandpaper lying flat on a surface plate quickly removes protrusions and leaves surfaces clean and flat.

To maintain the flattest surface on small parts, it is best to lightly sand the part in a figure eight pattern. By using the figure eight pattern you are constantly changing the leading edge of the part which is the area that usually gets sanded the most.

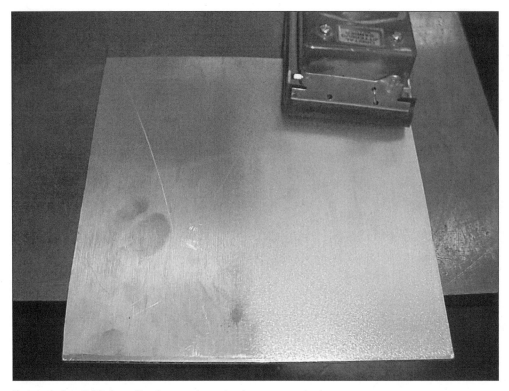

FIGURE 4–8   Orbital sanding is a great way to clean up raw stock.

FIGURE 4–9   A machined surface is "tuned up" over a sheet of 600 grit wet/dry sandpaper.

FIGURE 4-10 Parts that get handled a lot should have their sharp corners filed off.

12. *Use a tumbler for final deburring.*

    Tumblers work best for removing sharp corners, not large burrs. You can prepare parts for tumbling by filing surfaces flat which leaves sharp corners. Let the tumbler do the work of rounding the corners.

13. *File off the sharp points of parts that get handled a lot. (See Fig. 4-10)*

    Parts with dull corners are infinitely more pleasing to handle than sharp cornered parts.

14. *Countersink screw holes with dedicated countersink bits. (See Fig. 4-11)*

    These countersinks come in different sizes and are designed to cut correct countersink geometry for each of the different flathead screw sizes. They cut an upper diameter that is slightly larger then the maximum diameter of the corresponding screw head. These bits save cutting extra material and produce a clean countersunk hole.

FIGURE 4–11 Dedicated countersinks work great for cutting clean, on-size counter sunk holes. These cutters were made by Cleveland Twist Drill.

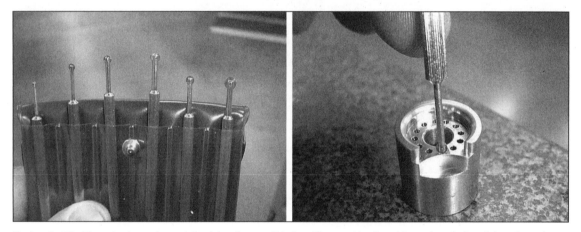

FIGURE 4–12 These tools work great for deburring small holes. They can be found in various industrial tool catalogs.

The pilots on these bits reduce chatter and when used in a drill press, help line up holes.

15. *Deburr tiny holes with the proper tools. (See Fig. 4-12)*

   With this set of hand tools you can easily deburr holes down to about .015″ in diameter. The tools are self-centering and cut with little pressure. They leave an appropriately dressed edge that looks clean and professional.

FIGURE 4–13 Live centers with protruding points provide better tool clearance when working near the tailstock.

16. *Drill small centers in lathe parts.*

   The bearing surface of a center only has to be large enough so that it doesn't deform under pressure. 3/32"–3/16" outside diameter centers would likely be sufficient for the majority of small tool room jobs.

17. *Use live centers. (See Fig. 4-13)*

   Unless you are turning at a snail's pace, dead centers have a tendency to heat up and gall in spite of whatever miracle grease you use.

   Live centers work well and there are large varieties to choose from. I prefer live centers with small protruding points for better tool clearance.

18. *Use your air hose liberally.*

   Blow long and hard. Try to get all the chips and residue out of your collets, vises and other clamping devices. If you don't remove all the chips, you'll likely end up pressing them into other parts.

19. *Set parts down gently.*

   Avoid throwing small parts into a box like you were shooting basketballs. They'll get nicked if you do.

20. *Wash parts with non-sudsing soap.*

   This is a good way to finish a job. Washing parts in soap and water removes dirt and oil and makes them look better. Chemsearch Swoop® skin cleaner works well for this purpose.

Help for Novices

What is the best way to learn something? For my money it's hard to beat the "just do it" method which may land you in hot water once in awhile. So be it. This chapter is about basic shop practices and pointing newcomers in the right direction. It's also about avoiding blunders. Not all mistakes are of the dimensional type in a machine shop. Blowing chips all over the guy next to you would be an example of a "non-dimensional" mistake that would likely irritate somebody. By learning some of these basics; you'll be in a better position to work independently and effectively in conjunction with other shop personnel.

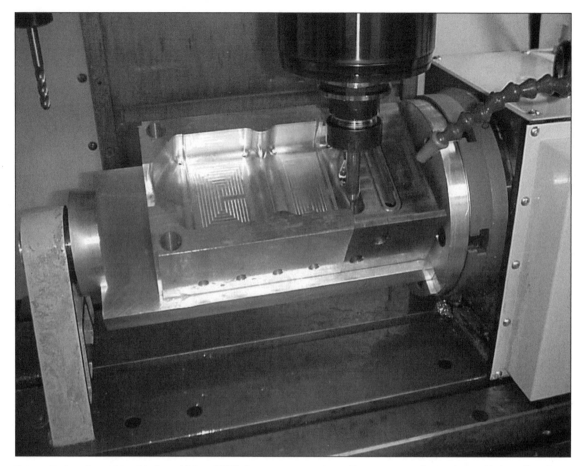

FIGURE 5–1   Computer aided machining (CAM) is now commonplace. Newcomers to the trade would do well to learn the ins and outs of conventional as well as CNC machining.

1. *Maintain sharp computer skills.*

    There's a lot of fancy machining going on these days. Computer aided machining is now common place and newcomers to the trade will essentially be obliged to learn CAD/CAM (computer aided design/computer aided machining) to be become more employable. I'm convinced, however, that machinists and craftsman with good conventional skills, good mechanical aptitude and an eye for detail will always be needed. Concentrate on improving both your computer and conventional skills as you progress to increase your value as a toolmaker.

2. *Learn to use your indicator.*

    One of your greatest assets as a machinist is going to be your ability to use an indicator to inspect and line things up. It is a skill that comes with practice so practice whenever you get the chance. One suggestion I can give you is that it is best to position the tip of your indicator on the surface being indicated using a small amount of preload. It is easy to get lost in needle rotations if you use too much preload. Also, a small

FIGURE 5–2   Deburring the edges of thin plates in a disk sander is very dangerous if the support table is used. I know of two people that severely ground down their thumbs doing this.

amount of preload allows you to move the tip on and off a feature with little risk of the tip getting knocked out of position.

3. *Do not wear gloves or jewelry around machinery.*

   Gloves and jewelry are killers. They can easily snag on rotating spindles, sanders and cutters and suck you in. Once in the work force, you may get bombarded with so many safety issues that ultimately you may not pay much attention to a lot of it. If I were to recommend just one safety rule to follow, it would be "DO NOT WEAR GLOVES WHEN WORKING AROUND MACHINERY."

4. *Avoid using the support table of a disk sander to deburr thin plates. (See Fig. 5-2)*

   I've worked with two people that severely ground down their thumbs trying to deburr the edges of a thin plate in a disk sander. As they were sanding an edge, the plate got sucked into the small opening between the table and the disk which pulled the plate

FIGURE 5-3   Bar stock that extends beyond the rear of a headstock can whip dangerously unless it is contained. It's a condition that must be avoided.

and their thumbs into the disk. If you are going to deburr a plate this way then make sure you hold the plate above the support table to do so.

5. *Before cutting raw stock to length, plan how to run the job.*

   Sometimes, especially on lathe jobs, you may need extra material or length to hold a part. Plan on how you are going to run the job before cutting raw stock to length to avoid limiting your options.

6. *Have backup material on hand whenever possible.*

   Most novices make more mistakes than experienced people. Nevertheless, we all make mistakes no matter how experienced we are. I've found that the stress of ruining a part is enormously reduced if you have another piece of material to start machining right away.

It is amazing how fast you can catch up under these circumstances. The dimensions will be fresh in your mind, you'll have your tooling ready and neither you nor anybody else will have much time for complaining if you're busy making chips.

If you don't have backup material to start machining right away, mistakes have a way of festering and getting blown out of proportion.

7. *Avoid using compressed air to remove fiberglass chips.*

   Fiberglass and other fibrous materials can be irritating to the skin. I dislike machining fiberglass simply because it is nearly impossible to keep the residue off your clothing and skin. Nevertheless, we must try. Therefore, never blow fiberglass chips with compressed air or you'll end up filling the shop with a nasty airborne irritant. Use a damp brush or paper towel and a dustpan.

8. *Avoid using a new band saw blade to cut fiberglass.*

   Cutting fiberglass with a standard band saw blade quickly ruins the blade. If you want to cut fiberglass in a band saw then you should use an old worn out blade to do so. You can also cut fiberglass with a milling machine using a small carbide end mill.

9. *Avoid extending bar stock much beyond the headstock of a lathe. (See Fig. 5-3)*

   Bar stock extended too far beyond the rear of a headstock can easily bend and whip even at moderate spindle speeds.

10. *Clear the base metal of a brazed carbide tool before using a diamond grinding wheel to sharpen the carbide. (See Fig. 5-4)*

    Grinding soft base material with a diamond grinding wheel will load up the wheel and prevent it from cutting.

11. *Avoid using a granite surface plate as an anvil.*

12. *Avoid using compressed air to blow off a grinding machine.*

    Use a brush instead so that you don't spread grit everywhere.

13. *Always pull a wrench when possible to avoid knuckle busting.*

14. *For increased rigidity put the quill of your mill all the way up when hogging.*

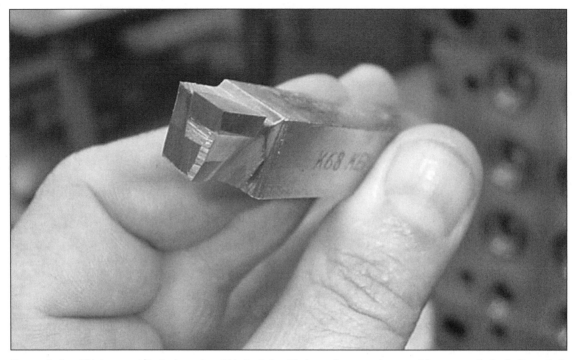

Figure 5–4   The base metal of a brazed carbide tool should always be ground away with a snag grinder before grinding the carbide with a fine diamond grinder.

15. *Never pound on gauge pins.*

16. *Use masking tape to help line up letter and number stamps. (See Fig. 5-5)*

17. *Use toilet paper to spread Dykem. (See Fig. 5-6)*

   I've seen various methods for applying and spreading Dykem® from brushes to sponges to sprays. The method I like best for applying Dykem to a large area is to use a wad of toilet paper to spread the Dykem after pouring a little blob on the part.

18. *Respect your hands.*

   Avoid using your hands to hammer with. Your hands are your most versatile and valuable too. Be careful with them.

19. *What's the difference between a jig and a fixture?*

   A jig is a guiding device and a fixture is a holding device.

FIGURE 5–5   A piece of masking tape is used as an aid to line up hand stamps.

20. *Is there an optimum speed for running a standard edge finder?*

Edge finders will work at just about any speed. Running them too fast, however, can cause the tip to whip and running them too slow won't give you a crisp reading. Consensus is that they should be run in a range from about 800 to 1200 RPM.

21. *What is "tooling"?*

It is a broad term used to describe any piece of hardware whether it is a cutter, jig, fixture, measuring device or accessory that aids in the holding or machining of a part.

22. *What are CNC machines?*

CNC stands for Computer Numerical Control. CNC machines are different from conventional machines in that cutter positions, speeds and feeds can be controlled with program code. With a two axis CNC mill only "X" and "Y" table positions can be controlled. With a three axis CNC mill the "X", "Y" and "Z" positions can be controlled. A three axis machine equipped with a programmable indexing head gives the machine a fourth axis. If the indexing head can be programmed to tilt, that would be considered a fifth axis. Another type of five axis machine is one with a tilting spindle.

FIGURE 5–6   A paper towel is used to spread Dykem®.

FIGURE 5–7   An indicator is used to line up a part for rework. A total indicator reading (TIR) is given when the part is spun which is equal to half the actual off center error.

23. *What is a "machining center?"*

It is a sophisticated machine tool capable of doing multiple machining operations in one location.

24. *What are nominal sizes?*

Nominal sizes are convenient labels used to discuss sizes that in reality nobody will ever hit. There is always some tolerance or error in any machined part.

You'll hear people say, "Just shoot for nominal". That means if the nominal size you're working to is .5″ then you need to do your best within reason to machine the part to a half-inch taking into account the type of machining process you're using. You wouldn't be expected to hold a half-inch nominal size as closely in a milling machine as you would in a surface grinder.

25. *Quick check material hardness with a file.*

Hardness is very important in all aspects of metalworking. To give you a feel for hardness, I've compiled a short list for your reference. The items tested in this list were all high quality brand name items.

ITEM	RC HARDNESS (Rockwell, C scale)
Socket Head Cap Screw	42
Hex wrench	50
6″ Scale	49
Screwdriver Blade	49
Locking Pliers	55
Hobby knife Blade	56
Height Gauge Scribe	58
Single Edge Razor Blade	58
Case Hardened Dowel Pin	59
File	65
Piece of High Speed Steel	65
Piece of "Micro Grain" Carbide	79

One way to find out if material is hard is by running a file over an edge or corner of the part. The file will slide over hard material without cutting it.

26. *What is "TIR"? (See Fig. 5-7)*

It stands for "Total Indicator Reading."

One thing a machinist must understand about "total indicator reading" is that when indicating any cylindrical surface relative to another cylindrical surface running dead

FIGURE 5–8   A drop indicator is used to check the depth of a narrow groove.

true, the axis of the indicated surface differs from the axis of the surface running dead true by half the amount of the total indicator reading.

Another abbreviation used for TIR is FIM which stands for "full indicator movement."

27. *Use a drop indicator for checking the depths of narrow hard to reach places. (See Fig. 5-8)*

28. *When taking inside measurements with large vernier calipers, be sure to read the correct scale. (See Fig. 5-9)*

The top scale on the calipers shown in the photo should be read when taking inside measurements with the protruding tips.

29. *Calculate sine plate stack-ups correctly.*

When calculating sine plate stack-ups, multiply the sine of the angle by five if you are using a five inch sine plate, ten if you are using a ten inch sine plate, twenty if you are

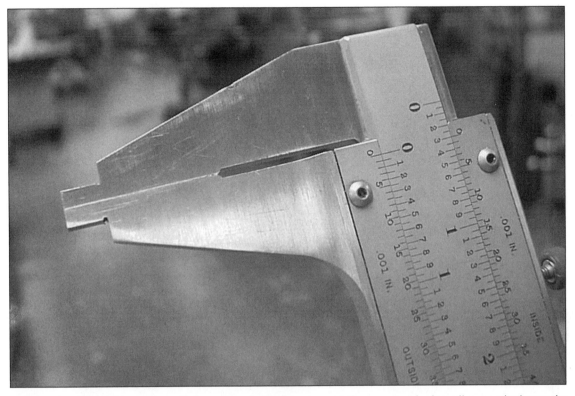

FIGURE 5–9 With large vernier calipers you must remember to read the proper scale depending on whether you're measuring an inside or outside dimension.

using a twenty inch sine plate and so on. It is a good idea to double check gauge block stack-ups with calipers.

30. *Divide metric dimensions by 25.4 to convert to English dimensions. (millimeters to inches)*

To convert metric dimensions (mm) to English dimensions (in) with a calculator, enter a metric value and divide by 25.4. To do more than one conversion, with most calculators you can just keep entering values and hitting the "equals" button to output an answer. In other words you don't have to keep entering 25.4 for each calculation since that value and the operation performed (division) are stored in the calculator's memory. The method also works for other types of calculations such as addition and subtraction as well.

31. *Understand the difference between "absolute" and "incremental."*

Absolute dimensions refer to distances from a single origin to a point or feature. Incremental dimensions refer to distances from one feature or point to another. Most digital readouts can be switched back and forth. One example where switching could be useful is if the centers of a series of bolt circle patterns are dimensioned from one

FIGURE 5–10 Moderate spindle speeds and high chip loads work well for machining plastics.

corner of a plate. The centers of the patterns could be dialed off in "absolute" then by switching to "incremental" the holes could be dialed off incrementally.

32. *Machine plexiglas and other plastics with slow spindle speeds to avoid melting them. (See Fig. 5-10)*

    Relatively slow spindle speeds and high feed rates work best when machining plastics. You want to avoid spindle speeds that create enough heat to melt the plastic.

33. *Use a drill press vise to hold round stock for cutting in a vertical band saw. (See Fig. 5-11)*

34. *In a section view, the direction of the arrows indicate the direction you would look to see the section.*

FIGURE 5–11 A small vise is used to hold bar stock for cutting off in a vertical band saw.

35. *Apply the 45-30-30 rule when filing edges.*

This rule means that when filing an edge, the majority of the filing you do should be done with the file tilted at a forty-five degree angle. After that you should tilt the file at a thirty degree angle from each face to clean up any remaining "fuzz".

36. *Break through material gently when drilling holes to avoid throwing large burrs.*

37. *Use an eight-inch mill bastard file for most filing and deburring.*

They are easy to clean, fit comfortably in your hand, remove stock quickly and cut smooth surfaces.

38. *Use a tear drop file for filing radii. (See Fig. 5-12)*

Tear drop files work well for filing radii because the shape of the file allows you to rotate the file and roughly match the radius being filed.

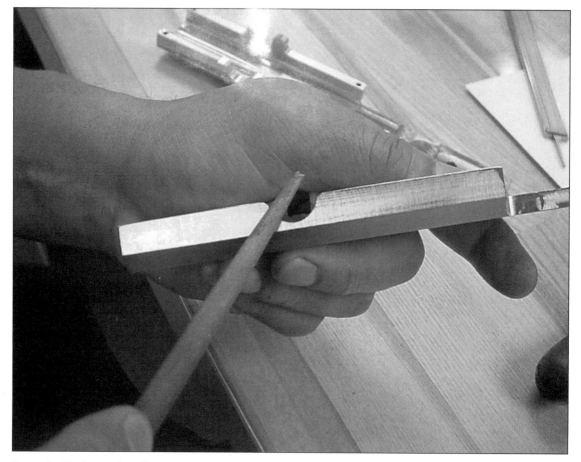

FIGURE 5–12 A tear drop file is used to deburr a curved shape. Tear drop files work well for deburring shapes with varying radii.

39. When should you use washers under tie-down clamp nuts?

Use washers under clamp nuts when you want to avoid the twisting and side loads put on clamps tightened down without them.

40. What does normal to a surface mean?

Normal to a surface means perpendicular to a surface.

41. Are there formulas for converting gauge sizes to dimensions?

No. A characteristic of nearly all gauge sizes is that the size is given by a number and you need to refer to a table to find exactly what size or dimension that number stands for.

FIGURE 5-13 A drill gauge is used to check the angle and symmetry of a drill bit.

42. *Use a drill guide to maintain symmetry when hand grinding drills. (See Fig. 5-13)*

43. *Make yourself some simple hand held deburring tools. (See Fig. 5-14)*

    I've been using these same three tools for decades.

44. *Don't force cutters.*

    Properly ground cutters cut. If you have to put undue force on a cutter something is wrong. Find out what it is and correct it before you proceed.

45. *Dull cutters throw large burrs.*

46. *Use the shortest, sturdiest cutters possible to enhance rigidity and improve machining efficiency. (See Fig. 5-15)*

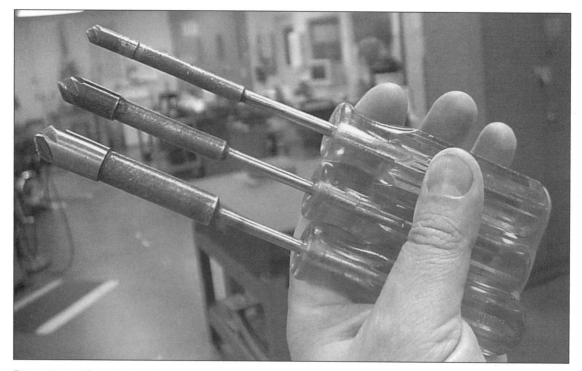

Figure 5–14 These hand held deburring tools can be used for various hole deburring tasks.

Figure 5–15 A corncob cutter is used to rough off material.

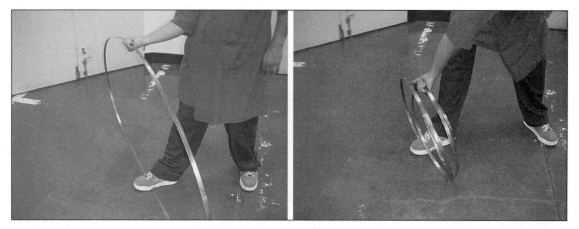

FIGURE 5–16 Folding a band saw blade is easy. Unfolding one without shredding yourself takes some skill.

47. *Aim to miss tolerances on the metal safe side.*

   It is always easier to take metal off than to put it on.

48. *Fold up a band saw blade the easy way. (See Fig. 5-16)*

   If you are right handed, step on the unfolded blade with your right foot. Hold the top of the blade, palm up in your right hand. Twist the blade as far as you can counter clockwise to fold the blade into three loops.

49. *Avoid using cheap edge finders.*

   Poor quality edge finders don't work very well and may give you significantly false readings. Starret® edge finders are high quality tools that provide consistently accurate readings when kept clean and lubricated.

50. *Use a magnet to find out if a material is stainless steel.*

   Many stainless steels are non-magnetic but not all of them. 300 series stainless steels are non-magnetic. 400 series stainless steels that can be heat treated are magnetic.

51. *What is the countersink angle of a metric flat head screw?* Ninety degrees.

52. *What is the countersink angle of a standard flat head screw?* Eighty-two degrees.

53. *What is the countersink angle of an aircraft flat head screw?* One-hundred degrees.

54. *What is the maximum shoulder fillet radius of a lathe part supposed to be if there is no callout on the drawing?*   .015″

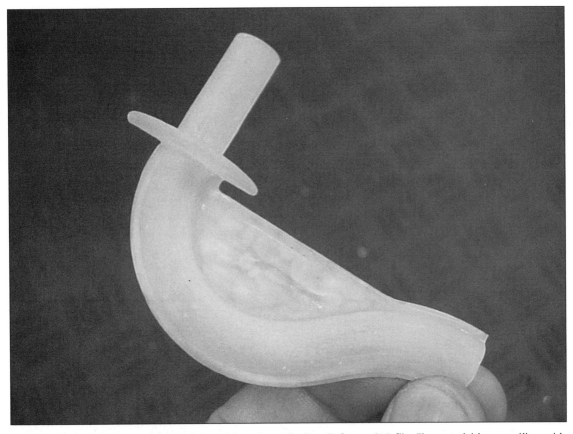

FIGURE 5–17 An SLA (stereolithography) model can be made directly from a CAD file. The material is epoxy like and is made with a machine that uses a laser to harden successive layers of material.

55. *What is a microinch?*

It is a unit of measure designated by the symbol "μin" and is equal to one millionth of an inch. A micron is designated by the symbol "μm" and is equal to one millionth of a meter.

56. *What does the symbol "Ra" refer to?*

The symbol means "roughness average." When called out as a surface finish, the number designates the average deviation from the mean line of the surgace texture commonly expressed in microinches.

57. *What is "Stereolithography"? (see Fig. 5-17)*

Stereolithography, or SLA creates a physical model of a part directly from a CAD file. An SLA is made with a machine that uses a laser to harden successive layers of a plastic type material to create a model. The process falls under the umbrella of "rapid prototyping".

FIGURE 5–18 The core in this mold is an example of a lofted surface created with CAD/CAM equipment.

58. *What are lofted surfaces? (see Fig. 5-18)*

Lofted surfaces are surfaces created by sweeping or blending different profiles from different planes. For example a lofted surface could be created by blending a square in one plane to a circle in another parallel plane.

59. *What is the American drawing format?*

The drawing format used in the U.S. is called "3rd angle projection".

60. *What is the European drawing format?*

The drawing format used in Europe is called "1st angle projection".

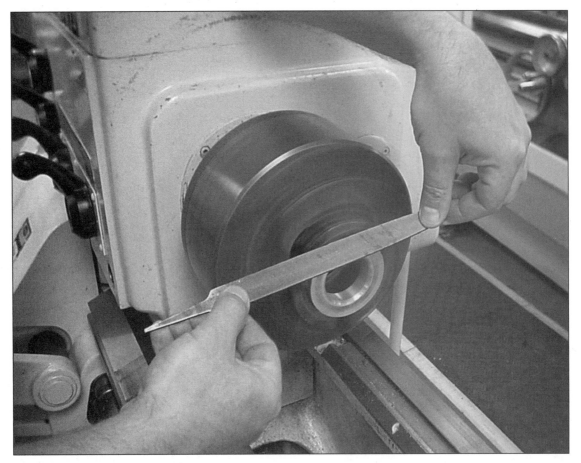

FIGURE 5–19 If you are going file a part in a lathe without using a file handle, make sure you position the tang of the file off to the side of your hand to avoid hand injury.

61. *Hold files carefully when filing parts in a lathe. (See Fig. 5-19)*

It's best to use a file handle to file parts in a lathe to avoid hand injury. If you don't use a file handle then you have to be especially careful about holding the tang of the file off to the side of your palm so that if the file runs into the chuck, it won't impale your hand.

62. *When making mold parts, be careful about what edges you file.*

I prefer leaving all edges sharp when making mold parts until later when I can carefully examine and detail them at a bench. Most machinists are so accustomed to filing edges they do it automatically. When making mold parts, filed edges can spell disaster especially on core and cavity details.

Parting lines, shutoffs and other corners and edges on mold parts may have to be perfectly sharp in order to prevent flash and maintain part integrity.

63. *Avoid generating too much cutting oil smoke.*

Nobody wants to breathe cutting oil smoke. Sometimes you can't avoid using cutting oil, like when you're tapping, cutting a groove or forming something in a lathe. It is also beneficial to use when trying to cut smooth finished surfaces. Nevertheless, I think some machinists, especially newcomers, overuse cutting oil.

64. *Slow down squealing cutters.*

A squealing cutter is an indication that you either need to slow down the spindle, increase feed or sharpen the cutter. There is an exception. Drill bits over about an inch in diameter tend to make a lot of noise when they cut, even at slow spindle speeds. Don't be shy about cranking in a large, sharp drill bit aggressively.

65. *Use HSS (high speed steel) or cobalt instead of carbide to machine aluminum.*

HSS usually leaves a better finish when cutting aluminum and has less tendency to accumulate a built up edge.

66. *Whenever you're having difficulty with a piece of tooling or running a job of some sort, sit down and do a thorough inspection of what you have.*

"Inspecting what you have" is a simple yet effective approach to begin troubleshooting. Many people approach problems haphazardly and therefore never get a handle on a problem. If you do a thorough inspection and record your results either by writing dimensions directly on parts or on a piece of paper, you'll be in a better position to take corrective action.

chapter **6 More Shop Talk**

One of the advantages of learning the machining trade is that much of the knowledge you learn is cumulative and enduring. Many of the techniques discussed in this book were valid decades ago. Likely much of this information will be useful decades from now.

The suggestions offered in this chapter are going to include more information about common and possibly not so common shop practices. Some of these suggestions may seem obvious. I've always found that "obvious" hardly ever is. Ask ten different people a simple question like "What is a cone shaped edge finder used for"? You'll be amazed at the variety of responses you'll get.

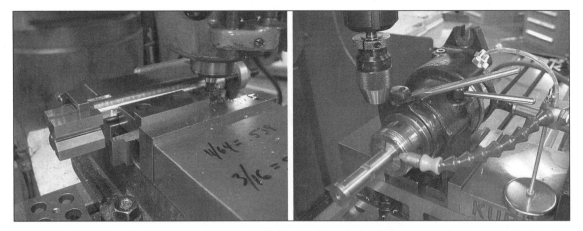

FIGURE 6–1   Indexing parts for machining takes a little creativity. These photos show a few methods for indexing parts conventionally.

FIGURE 6–2   CNC indexing heads are easy to use and can save a substantial amount of labor.

FIGURE 6-3 Clamping over "air" is almost always a bad idea. In this case the strap clamp could damage the vise if it were tightened down too tightly.

1. *Get creative when indexing parts. (See Fig. 6-1)*

   There are many ways to cut features relative to one another. There is no right or wrong way as long as your setup is rigid and accurate. One way is to use an indexing head. Another way is to use a V-block in combination with a machinist's square. Another way is to index on an existing feature and so on.

   CNC indexing heads are easy to program and can save a substantial amount of time. (See Fig. 6-2)

2. *Avoid clamping over air. (See Fig. 6-3)*

   I used to work for a grumpy old German toolmaker that knew the trade well. In one instance the German guy loaned his precision grinding vise to one of the younger guys in the shop. The younger guy was setting up a job and clamped the vise down on the table as shown in the photo.

Figure 6-4   Four jaw chucks are the best way to hold round stock when there is little stock to hold.

When the German guy saw how his vise had been clamped with the pressure of the clamp being applied over the middle of the slot in the vise, he darn near had a heart attack.

The moral of the story is that you should try to avoid clamping over anything that does not have solid support directly underneath it.

3. *Use a four jaw chuck when you don't have much stock to hold. (See Fig. 6-4)*

    An independent four jaw chuck will hold parts more securely than a three jaw chuck especially when there is minimum stock to grip.

    With a little practice you'll be able to line up parts in a four jaw chuck in no time.

4. *Remove Dykem® from parts before sending them to heat treat.*

    I heard a story once about some parts that were sent to heat treat with dried Dykem on them. The story goes that when the parts came back they were extremely warped.

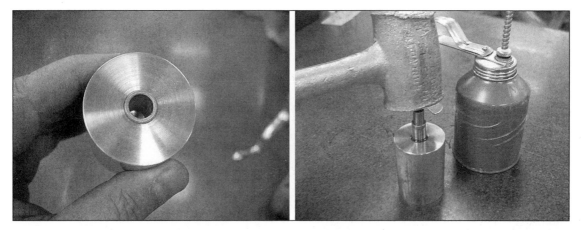

FIGURE 6-5   A bushing is removed from a blind hole by using hydraulics.

When the heat treat company was questioned as to why the parts were so warped they responded that it was because of the dried Dykem that was left on them.

To be safe and to eliminate one possible excuse from the heat treat people for warped parts, it's a good idea to remove Dykem before sending them out.

5. *Use hydraulics to remove bushings pressed into blind holes. (See Fig. 6-5)*

How can you extract a pressed in bushing or dowel from a blind hole? It can be difficult unless some provision was made beforehand.

One technique that works well is to use hydraulics.

I remember once when I was setting up to use this technique, my lead man took notice and thought I was wasting my time. He had no clue what I was doing as I was preparing to knock out some bushings from blind holes in a mold base. When the bushings popped out without a hitch, he could hardly believe his eyes.

This technique can only be used when there is a hole all the way through the axis of the bushing or dowel pin. Ideally dowel pins shouldn't be pressed into blind holes unless there is some provision for extracting them.

The way to use hydraulics to extract a bushing from a blind hole is to first fill the ID of the dowel or bushing with oil or grease. Any oil will do but thicker oil tends to work better because it doesn't run or splatter so easily. Then insert a close fitting pin into the ID of the dowel or bushing. When you rap the end of the pin with a hammer, you force oil around the end of the bushing which exerts the backward pressure needed to extract it.

6. *Use a spring loaded stop in a cutoff saw. (See Fig. 6-6)*

A spring loaded stop allows parts a little free movement when the blade cuts through. When a part is cut through using a solid stop there is a tendency for the part to get

FIGURE 6–6   A spring loaded stop used in a horizontal band saw allows the part to move a little when it is cut through so it won't jam the blade.

jammed between the blade and the stop which may damage the blade or pull it off the machine's drive wheels.

7. *Make a shallow cut first when using thin slitting saws. (See Fig. 6-7)*

   To keep thin slitting saws from walking off location on deep cuts it is best to make a shallow cut first about .05″ deep on the periphery of the part so that the blade has a track to follow when making a deeper cut.

8. *Close down oversize holes and diameters on lathe parts with a collet closer.*

   This is a great way to cheat. If you have a bushing or round part that has a diameter or ID that is slightly oversize, you can collapse the part with a collet closer as long as the wall thickness of the part is not too great.

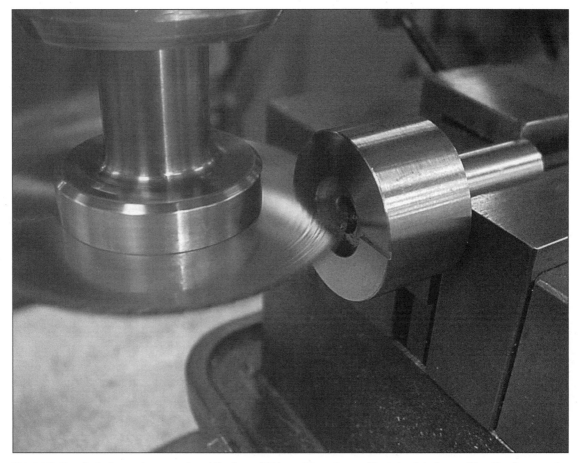

FIGURE 6–7   A shallow cut was made on the face of this part to act as guide or track before making the deeper cut.

9. *Run special cutters slowly.*

    If you have any kind of special cutter, form tool, or expensive cutter you would like to preserve then it is best to run them slowly. Metal cutting tools will cut at low RPMs.

10. *Use carbide on fiberglass with slow spindle speeds.*

    Fiberglass and other fibrous materials are abrasive and will quickly dull high speed steel cutters. Carbide holds an edge fairly well when cutting fiberglass and is generally the best choice for cutting abrasive materials. Use a hard, straight grade carbide such as C8 to prolong cutter life.

    Cut fiberglass with slow spindle speeds to keep from throwing fiberglass dust in the air.

11. *Stop the pentagon.*

    Once in a while a reamer will start to cut an irritating pentagon shaped hole which is undersized on the flats and oversize on its vertices. The pentagon results from the high

FIGURE 6–8   A reference cut was calculated and made to land on an angle. Once the part is set at the proper angle the reference cut can be cleaned up to precisely locate the face of the angle.

tool pressures created by using a dull, straight fluted reamer. Once the pentagon shape starts it can be difficult to stop. One way to stop it is by switching to a spiral fluted reamer otherwise you may have to bore the hole.

12. *Don't buy a lathe chuck that can't be adjusted for concentricity.*

Lathe chucks that can't be adjusted for concentricity are a pain. The only way to get something to run true in cheap, non-adjustable three jaw chucks is to bore soft jaws.

13. *Use a reference cut to precisely machine an angle. (See Fig. 6-8)*

One way to precisely machine an angle to the proper depth and location on a part is to first make a square reference cut. The sharp corner of the cut can be used as a reference mark to machine to once the part has been clamped in the vise at the proper angle. The photo shows a square reference cut on the lower right hand side of the block and what's left of a reference cut on the upper side which is in the process of being cleaned up.

FIGURE 6–9   A six inch scale is used to set a lathe tool on center.

The less accurate and more common way of machining an angle on a block is to machine to a scribed line on the block where the angle is intended to run out.

When using the reference cut method, it is imperative to use a sharp cornered cutter for maximum accuracy. The dimensions for the reference cut must be trigged out in such a way that the corner of the cut lands somewhere on the finished angle.

14. *Set lathe tools on center with a scale. (See Fig. 6-9)*

One way to set a lathe tool on center is by lightly pressing a scale between the point of the tool and the workpiece. An off center tool will tilt the scale. You can get a tool reasonably close to center by adjusting the height of the tool until the scale stands vertically. The only way I know to get a lathe tool exactly on center is to take light facing cuts and adjust the height of the cutter until it cuts through the exact center of the part.

FIGURE 6–10 A piece of brass is used to push chips and gunk out of a file.

15. *Clean files with a piece of brass. (See Fig. 6-10)*

Loaded up files don't cut very well and may leave scratch marks on smooth surfaces. One way to remove gunked up chips from a file is to use a piece of brass to push them out. This technique is a little time consuming but it works well on single cut files which are the files I prefer. Another way to get more life out of old clogged files is to sand blast them.

16. *Use an old file to chamfer the corners of lathe parts.*

Filing chamfers on lathe parts is hard on files. Even with the so called "lathe files" there is a tendency for the file to load up quickly. Use an old beat up file for roughing chamfers then switch to a clean, smooth file to finish them.

17. *Support keyway broaches up high. (See Fig. 6-11)*

Keyway broaches pushed through parts in an arbor press have an irritating tendency to dig in towards the lower end of the slot. Support the broach relatively high above the part with the guide bushing to eliminate that problem.

FIGURE 6-11 The guide bushing for this keyway broach is supported well above the part which helps keep the broach from tilting and digging in.

18. *Rough out large keyway slots with an end mill before broaching.*

The more material you remove from a keyway slot before broaching, the less force it takes to push the broach through. One way to remove material is by milling it away. Another way is to use a smaller broach first.

19. *Use prick punch marks to identify part orientation in assemblies. (See Fig. 6-12)*

20. *Use colored tape to identify raw stock. (See Fig. 6-13)*

Make a fixture to mount rolls of colored tape so that pieces of tape can be cut off and wrapped around stock to identify it. That way you'll never have to worry about marking pens running dry or getting lost. It is one of the most efficient systems I have used.

FIGURE 6–12 Punch marks can be used as visual aids to locate parts in assemblies.

FIGURE 6–13 Colored tape can be used to quickly tag and identify raw stock.

21. *Use good quality bench vises with smooth jaws. (See Fig. 6-14)*

There is something about having good quality bench vises in a shop that speaks well for the shop. I dislike using crummy bench vises and most other cheap tools for that matter.

FIGURE 6–14 A high quality bench vise is used to hold a part for hand tapping.

22. *Make lathe parts in the lathe and mill parts in the mill.*

   CNC equipment can make us lazy. You'll occasionally be tempted to make round parts in a CNC mill since it runs by itself. I've made lathe parts myself in the CNC mill but it is not something you should get in the habit of doing. By and large, lathe parts should be made in a lathe and mill parts in a mill.

23. *Remove as little material as possible when modifying soft jaws. (See fig 6-15)*

   Choosing the right size metal ring to clamp in your lathe chuck allows you to remove a minimum amount of material to arrive at a usable diameter. By removing a minimum amount of material you can extend the life of a set of soft jaws. Ring sets are available in most industrial tool catalogs.

24. *Install a shop rag on your conventional milling machine to stop flying chips. (See Fig. 6-16)*

   Staple a shop rag onto a piece of dovetailed wood and slide it onto the dovetailed portion of the ram of your Bridgeport. The rag keeps chips from ricocheting off the body of the mill. This arrangement is especially useful when fly cutting.

FIGURE 6-15 Choosing the proper size slug or metal ring for boring a set of soft jaws can extend the usable life of a set.

25. *Be careful when using pinch type quill stops. (See Fig. 6-17)*

   If you pull the quill down hard against a pinch type quill stop, the stop may open and slip down. I prefer other types of rapid adjusting quill stops that are unable to slip under pressure.

26. *Use a cutter whose diameter is slightly greater than the width of the part.*

   In a milling machine, it is not good practice to use a cutter that is too much greater in diameter than the width of the surface being machined. A large diameter cutter moving across a narrow workpiece has a tendency to "hammer" the workpiece which can shorten the life of the cutter. The hammering is caused by the cutter entering the material at or near its maximum chip load; whereas a cutter that is slightly wider than the workpiece, enters the material "early" and gradually increases the chip load as it rotates into the workpiece then gradually decreases the chip load as it leaves the workpiece.

FIGURE 6–16 A shop rag is used to keep chips from ricocheting off the body of the mill.

The chip thickness is greater in the first half of the chip because of the lateral movement or feed of the table, which makes the first half of the chip curl more than the last half. That's why some chips look like the number six.

The gradual increase and decrease in chip load on a cutter that just covers the workpiece makes for a smoother more efficient cut.

If you want to avoid the hassle of changing from a large diameter cutter to a small diameter cutter, then the next best thing to do is position the cutter nearly tangent to the workpiece so that you'll be cutting the longest chip possible. Be sure to position the cutter on the correct side of the workpiece so that the cutter enters the material at minimum chip load. (See Fig. 6-18)

FIGURE 6–17 A pinch type quill stop is shown being used here.

27. *Use a Dremel® motor with a small mounted stone to mark and write on metal. (See Fig. 6-19)*

   The right hand photo shows the mounted stone being dressed with a Norbide® stone.

28. *Use a .001" shim to touch off a non-rotating end mill in "Z" to avoid gouging the work-piece. (See Fig. 6-20)*

29. *Avoid using glue to attach shims.*

   I've never found a really good way to permanently attach shims. Glue not only doesn't hold very well it creates thickness inconsistencies. I've seen people try to spot weld shims in place with inconsistent results. The spot welding may blow holes in the shims or create craters or not hold very well. For lack of a better way to hold shims in place, I use a thin layer of sticky grease.

FIGURE 6–18 When using a cutter that is a lot wider than the surface being cut, the cutter should be positioned to cut the longest chip possible.

FIGURE 6–19 A mounted stone is used to write on metal. The photo on the right shows the stone being dressed with a Norbide® stone.

FIGURE 6–20 An end mill is set to "Z" zero using shim stock.

FIGURE 6–21 A piece of material is clamped opposite the partial hole so the ID can be measured.

FIGURE 6–2 A gear is held with soft jaws to avoid damaging the gear teeth.

30. *To bore and measure a partial hole, clamp a piece of material opposite the partial hole. (See Fig. 6-21)*

31. *Use solid carbide boring bars to bore deep holes.*

    Hole diameter is a limiting factor when it comes to choosing a boring bar. To bore a deep, small diameter hole you'll sometimes be forced to extend a boring bar quite far beyond the tool holder.

    It is best to use solid carbide boring bars when boring deep, small diameter holes because carbide is more rigid than HSS.

32. *When boring the hub of a gear, use the gear teeth for reference. (See Fig. 6-22)*

    Gears need to run true in order to run smoothly. Ideally, the exact theoretical pitch diameter of a gear would be the best thing to hold for boring the ID or hub of a gear. Since that is not practical, the next best thing is to hold the OD of the gear. Aluminum soft jaws work well for holding gears that need hub machining. Soft jaws exert clamping pressure over a large surface area and are less prone to damage gear teeth.

FIGURE 6–23 An adjustable wrench is used to loosen the chuck from the spindle of this Hardinge lathe.

33. *In a Hardinge® type lathe, use an adjustable wrench to torque the chuck on and off. (See Fig. 6-23)*

34. *Use a mill in a pinch to make lathe parts. (See Fig.6-24)*

     When the lathes are tied up and you need to make a simple lathe part, do it in the mill by mounting the lathe tool in the vise.

35. *Use strap clamps in combination with your milling machine vise. (See Fig. 6-25)*

     Once in awhile you'll have a cut that can be quickly made using this setup.

FIGURE 6–24 You can use a milling machine in a pinch to make simple lathe parts.

36. *Use cutting oil to cut plastic.*

   Plastic has a tendency to spring away from cutting tools. The tendency is sometimes quite noticeable especially when tapping. The friction and clamping action of plastic usually requires increasingly more torque to wind in a tap the deeper you go.

   Cutting oil can reduce friction and significantly improve the machining characteristics of plastic.

37. *Sand blast the tapered shanks of your lathe tools for a more secure fit in your old lathe tailstock.*

FIGURE 6–25 The block clamped in the vise allows strap clamps to be used to hold parts on the vise itself.

38. *Use Scotch-Brite® to remove rust and other contaminants. (See Fig. 6-26)*

Scotch-Brite® has multiple uses in a machine shop. It can be used to remove rust, dirt and heat discoloration. It can also be used to smooth and hide blemishes on machined surfaces.

39. *Use the middle screw of your Criterion® boring head for locking and unlocking the slide. (See Fig. 6-27)*

The outer two screws should be used to adjust the slide tension. Once adjusted, the outer two screws can be left alone.

FIGURE 6–26 Scotch-Brite® is being used to remove discoloration from this core pin.

FIGURE 6–27 The center screw of a Criterion® boring head should be used to lock and unlock the slide. The outer two screws are used to adjust the tension on the slide. Once adjusted, the outer two screws can be left alone.

FIGURE 6–28 A thin layer of cutting oil is applied to this lathe part in preparation for a finishing cut.

FIGURE 6–29 Work stops should be constructed with minimal contact area to avoid trapping chips.

FIGURE 6-30 You can get more rigidity out of a lathe boring bar by angling it as much as possible.

40. *Use water mist for roughing and cutting oil for finishing. (See Fig. 6-28)*

Water mist and soluble oil coolants help dissipate heat but they don't do much to improve surface finish or reduce friction. Cutting oil works the other way around. It doesn't do much to dissipate heat but it works well for improving surface finish and reducing friction. To cut a smooth surface, apply a thin layer of cutting oil on the surface of a part before finishing it.

41. *Use stops with minimal surface contact area. (See Fig. 6-29)*

Stops are something machinists use often to accurately and repeatedly locate parts in machine tools. Stops should be constructed or set up in such a way that there is minimal surface contact between the stop and the workpiece. By doing so, there is less tendency to trap chips between the stop and the workpiece

Spherical shapes work well as stops because they don't trap chips and have little tendency to dent parts. They can also be set at angles relative to the workpiece. Construct stops rigidly so they don't move if bumped a little hard.

42. *Set boring bars at an angle in a lathe for increased rigidity. (See Fig. 6-30)*

One way to get more rigidity out of a boring bar used in a lathe is to angle the boring bar as much as possible or as much as the inside diameter and depth of the bore will permit.

FIGURE 6–31 To eliminate chatter, slow spindle speeds and a lot of cutting oil should be used to cut large forms such as this fillet.

43. *In a lathe, use slow spindle speeds to form shoulders with large fillet radii. (See Fig. 6-31)*

    Cutters that cut over a wide area have a tendency to chatter. This is often the case when cutting large shoulder radii with form tools. Reduce spindle speed down to a crawl when necessary to eliminate chatter. Use plenty of cutting oil.

44. *Be aware that three flute end mills are difficult to measure.*

    I've always shied away from buying three flute end mills simply because their cutting diameters are difficult to measure. You can't measure them with conventional mikes or calipers since they don't have opposing flutes.

45. *When should you use brass shims instead of steel shims?*

    Brass shims are easier to cut than steel shims. Brass shims disintegrate more quickly than steel shims under extreme pressures such as those encountered in molding machines.

FIGURE 6–32 A single flute end mill can be used to side mill flimsy plates to eliminate the lifting forces generated by helical end mills.

46. *How tight should you tighten a band saw blade?*

The answer is no tighter than necessary to keep the blade from walking off location. An over tightened blade does nothing to improve cutting efficiency and only serves to shorten the life of the blade by putting undue stress on the blade and weld.

47. *Use single flute cutters in a milling machine to avoid lifting parts. (See Fig. 6-32)*

Helical end mills have a tendency to lift parts. When cutting thin, flimsy parts the lifting action can be detrimental. Keep parts from lifting by using a straight flute cutter such as those that can be made with a single flute cutter grinder.

48. *Make a 45° fly cutter. (See Fig. 6-33)*

A steep angled tool in a fly cutter automatically gives the cutting point a lot of bottom clearance so that it cuts with less pressure. Commercially available fly cutters have such

FIGURE 6–33 A fly cutter such as this one with the bit mounted at a steep angle provides better tip clearance and more rigidity than an equivalent size shallow angled fly cutter.

a low angle there is a tendency for the bottom side of the cutter to rub which can induce chatter.

49. *Put a piece of folded up paper behind a stack of parts clamped edgewise in a vise to hold parts securely. (See Fig. 6-34)*

Occasionally you can save time by stacking parts in a vise and machining holes and pockets in one setup. Because of slight variations in part width you have to be careful that all parts are held securely. One way to do that is by clamping the parts together with a clamp. Another method that works well is to fold a piece of paper a few times and place the gathering between the vise jaw and the stack. When the vise is closed the paper will crush and make up for slight variations in part width. Put the paper on the movable side of the vise.

FIGURE 6–34 A gathering of paper is placed between a stack of parts and the vise jaw to make up for slight differences in part width.

50. *Sandblast surfaces for better bonding with epoxies and resins.*

Sandblasted surfaces create a better bond than smooth surfaces with epoxy, body putty, and other resins.

51. *Take a class in GDT (Geometric Dimensioning and Tolerancing) then hope you don't see too much of it.*

There are some benefits of GDT, some of which have been thoughtfully adopted and put into use.

One aspect I especially like about GDT is the way that any feature can be labeled and then controlled in a variety of ways with respect to some other feature on the drawing. That aspect of the system is both useful and easy to understand.

The problem with the system in my opinion is that it can sometimes be difficult to decipher; especially when the draftsman didn't make the callout right in the first place. This business of varying tolerances based on material condition (MMC or LMC) can get complicated quickly especially on things like bolt patterns and related features.

To be effective machinists have to make chips. They can't afford to spend time trying to decipher complex and erroneous GDT callouts.

Years ago I took a course in GDT. At the end of the course the class was asked to take a proficiency test on the subject. About half the people in that class failed the test and that was when the subject matter was fresh in their minds. I can only imagine what percentage of people would have passed that test a year later.

Proponents argue that GDT provides more room for error than coordinate tolerancing and therefore has the potential to save parts that would have been otherwise destined for the scrap heap. That is a valid argument. In practical terms though I think it's hard to beat the simplicity of standard coordinate tolerancing.

Systems, regardless of what they are, get adopted not shoved down people's throats. People eventually end up adopting systems, parts of systems or practices that are easy to use and produce results.

Fortunately, some of the more user friendly and useful features of GDT have been generally adopted and put into use.

chapter **7** **The Cutter Caper**

The essence of skillful cutter grinding amounts to being able to produce cutters that are sharp yet sturdy, cut with little pressure, and evacuate chips well. When machining problems arise, often the problem can be traced to a faulty cutter. If you know how to choose, grind, alter and inspect cutters, you'll be able to do virtually any job that comes along.

Cutter grinding is an area of our trade that I believe separates the craftsman from the hack. Paradoxically, cutter geometry is not the complicated issue many cutter and insert manufacturers would have you believe. There is likely a wide range of cutter geometries and grades that will work for any given job.

I remember asking an accomplished machinist some years ago what shape tip he liked to use for fly cutting. His answer surprised me and yet thinking back on the exchange, I realize how accurate his answer was. He said "anything." And he was right as long as you add a

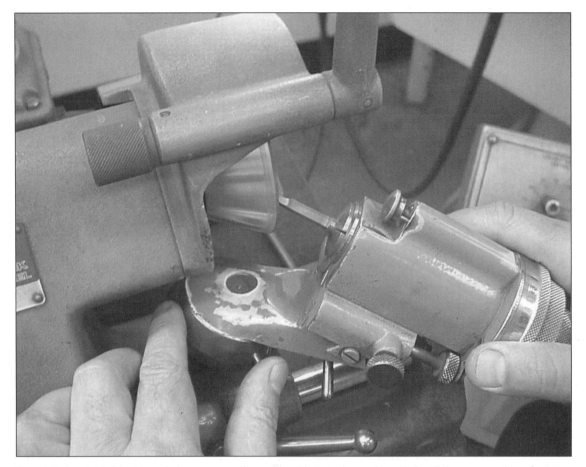

FIGURE 7–1   Deckel® cutter grinders are versatile machines that can be used to grind and sharpen a variety of cutters.

qualification. The leading edge of the cutter must be the first and only thing to contact the work. As long as that is true, just about any shape will cut.

Often, not having the right size cutter can stop a job cold. If you are able to make, sharpen or modify an existing cutter then you're time and money ahead. That way you won't have to deal with the people, the cost, the red tape and the delay involved in making an outside purchase.

There are many cutters you can make and sharpen with an ordinary pedestal grinder and/or surface grinder. It's best to have at least an end mill sharpener and single flute cutter grinder to be less dependent on outside sources for your cutters.

Many cutters can be made or sharpened by offhand grinding. You can use an ordinary bench or pedestal grinder to touch up most lathe tools, boring bars, drill bits, and fly cutter bits.

To make small precision cutters with angles and radii, single flute cutter grinders work well. (See Fig. 7-1 and 7-2) These grinders are small, easy to operate and quite versatile. They can be used to make a variety of cutters and can also be used to sharpen the bottoms of helical end mills.

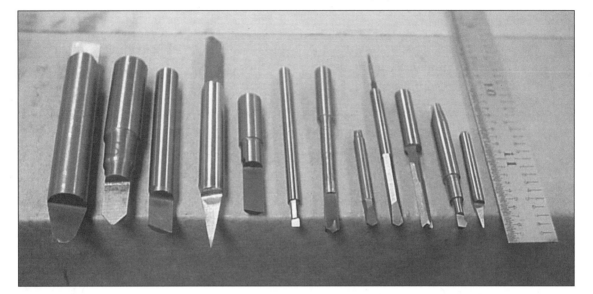

FIGURE 7-2   This photo shows a sampling of cutters that can be made with a single flute cutter grinder.

The types of machining you do and the products you produce usually dictate the type of cutters you prefer. Production machinists and mega metal movers tend to use insert type tools whereas machinists doing fine, close tolerance work often use brazed carbide tools.

I like to use brazed carbide tools for doing precision work in small lathes and insert type tools in larger lathes. Brazed carbide tools are easy to grind on a diamond grinder and can be sharpened and honed to cut with less pressure than off the shelf inserts. They are also readily available. I have a large assortment of brazed carbide and HSS tools ground in various shapes that are ready to go when the need arises.

The changeable insert type tools are usually better for roughing and doing work on rigid workpieces. Nevertheless insert type tools are commonly used in CNC machines for roughing and finishing. Inserts in the softer grade carbides like C2 thru C5 can take a lot of abuse without breaking or chipping. I prefer using softer grade carbides most of the time. They are less prone to chipping and in my experience, wear nearly as well as the harder grades.

Newcomers to the trade may hear the term "soft grade carbide" and conclude that softer grade carbides are right next to butter in terms of hardness. Not true; the soft grade carbides are plenty hard for cutting darn near anything.

One of the nice things about using insert type tools is that when the inserts do finally wear out they are easy to change and since they index into the same place, your settings won't change much.

In this chapter I'm not going to attempt to give any detailed instructions for using cutter grinders. The easiest way to learn the ins and outs of any machine is to have someone that is familiar with the machine go over it with you.

The following list of suggestions may help you sharpen your cutter grinding skills as well as your machining skills. Being able to make, choose and sharpen cutters will help you work efficiently and independently.

1. *Use an existing cutter as a guide for making and sharpening similar cutters.*

   Having an existing cutter to compare to can aid in setting up a grinder and choosing rake, relief and clearance angles.

2. *Grind cutters with a few degrees positive rake when possible so they cut with less pressure.*

   The rake angle of a cutting tool is the angle the chip slides over. The relief angles of a cutting tool determine the acuteness of the cutting edge. Clearance angles provide clearance between the tool and the workpiece. I prefer positive rake cutting tools most of the time because they tend to cut cleaner surfaces with less pressure. About five degrees usually does the trick. Too much rake and relief needlessly weakens a cutting edge.

   To give the reader a point of reference for grinding lathe tools I'd like to point out that many factory ground right and left hand brazed carbide lathe tools have rake, relief and clearance angles of about five degrees. That grind, in my opinion, is better for roughing than finishing. I prefer a little more relief and clearance on lathe tools used for finishing.

3. *Shy away from buying and using negative rake tools. (See Fig. 7-3)*

   I can think of a few reasons why you might want to use negative rake tools and those reasons hold true mainly for insert type tools. With a negative rake tool holder you can use inserts with zero relief which allows both sides of the insert to be used instead of just one. In other words, you'll get twice the number of cutting edges out of inserts designed for use in negative rake tool holders. Also one could make the argument that zero relief inserts are stronger than relieved or positive rake inserts.

   There is a trade off. Negative rake tools cut with more pressure than positive rake tools and build up more friction heat than positive rake tools.

4. *Grind HSS (High-Speed Steel) fast and hard.*

   You may have heard that while grinding HSS you shouldn't let it get too hot or it will lose its hardness. I've always ground HSS aggressively and never noticed a reduction in its hardness.

   Just to remove any doubt, I did a test one day. I took an unaltered piece of HSS and checked it for hardness. It measured about 65 RC which is about what you'd expect. Then I went to our old worn out pedestal grinder and abused the heck out of that piece of steel. I got it red hot as I held it mercilessly against our old glazed over grinding wheel. Then without dipping it in water, I checked the hardness of the face I had just ground. It turned out to be as hard as it was originally.

   Red hardness or hot hardness are terms used to describe a steels ability to resist changes in hardness at elevated temperatures. High speed steel alloys such as M42

FIGURE 7–3   Inserts used in negative rake cutters can be flipped over to provide twice as many cutting edges as positive rake inserts. The trade off is that negative rake cutters cut with more pressure than positive rake cutters.

which contain cobalt have higher hot hardness values than alloys such as M2 that don't contain cobalt.

5. *Web thin drill bits and center drills so they cut with less pressure. (See Fig. 7-4)*

Anything that decreases cutting pressure without sacrificing anything else in my opinion is a good thing. Web thinned drill bits and center drills will cut with less pressure if you do the web thinning properly. Although there are cutter grinders that sharpen and web thin drill bits using special fixturing, I have always sharpened and web thinned drill bits by hand.

Depending on the size of the drill, I may do the web thinning in either a pedestal grinder, which usually has a relatively large corner radius worn into the wheel, or by using whatever wheel may be mounted in the surface or cutter grinder at the time.

There are a couple of things you should be aware of when web thinning. First, you don't want to make the tip of the drill too weak. Reducing the thickness of the web (the

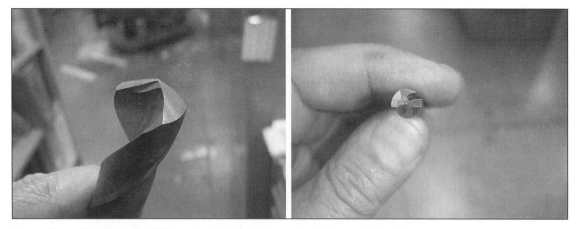

FIGURE 7–4   Web thinned drill bits cut with less pressure which helps maintain hole size.

material remaining between the flutes of a drill bit) by about seventy percent is usually adequate to substantially reduce cutting pressure. Second, it's best to grind the web area with a little positive rake so chips slide through the area with less difficulty.

I like to think of web thinning as a process of extending the pre-existing flutes. Blending the extension or web thinned area into the pre-existing flutes while maintaining a little positive rake works well. For drill bits to cut close to size, it's best to maintain as much symmetry as possible in all aspects of drill grinding including web thinning.

6. *Use progressively more relief angle with progressively smaller diameter cutters and end mills. (See Fig. 7-5)*

Clearance, rake and relief angles are not an exact science.

The only time I use numbers to grind cutters is when I have to choose a number in order to set a fixture or something. I use a drill gauge to grind drill bits but that is mainly to check symmetry.

To more skillfully grind rake and relief angles on all your cutting tools there are a few things you need to consider such as:

• The kind of material the cutter is going to be used for. I don't like to overly

FIGURE 7–5   Single flute carbide cutters work well for cutting hard and abrasive materials. Typically the smaller the cutter, the more relief angle needed to avoid rubbing.

complicate the issue. I basically want to know if the cutter is going to be used to cut tough material or easily machined material.

- Is the tool going to be used mainly as a roughing tool or a finishing tool? Roughing cutters and cutters used to cut tough material are generally ground with less rake, relief and clearance simply because they need to be stronger and wear longer than finishing tools. The more acute edge of a finishing tool won't last as long but will allow the cutter to cut with less pressure and friction which gives a better finish and more accurate cut.

- To what diameter, if applicable, is the tool going to be ground? The main reason for concern with relief angles on regular and single flute end mills is that you don't want any part of the tool behind the cutting edge to rub. Typically the smaller the end mill the more relief needed. I like to use about fifteen degrees on single flute cutters in the 5/16″–1/2″ diameter range, twenty degrees on cutters in the 1/8″–5/16″ diameter range, and thirty degrees on cutters smaller than about 1/8″ in diameter.

Typically single flute cutters are split down the middle which by default creates a zero degree rake angle.

7. *Make small six sided milling cutters for maximum sturdiness and ease of sizing. (See Fig. 7-6)*

Machinists need small end mills of various sizes to mill slots, O-ring grooves and other features.

I often prefer six sided single flute end mills over helical end mills. I like them for the following reasons:

- I believe from experience that they are the strongest cutters in terms of being able to take side loads without breaking.

- They're easy to make. You can quickly make six sided single flute end mills either in a surface grinder using an indexing head or in a single flute cutter grinder.

- When the bottom of a single flute cutter gets dull,

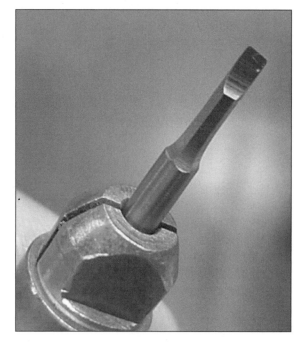

FIGURE 7–6   These little hex bits are strong, easy to grind and cut smooth surfaces.

FIGURE 7–7   Square shanked boring bars can be mounted in lathe tool holders as well as boring heads.

it can easily be resharpened by plunging the bottom of the cutter into a grinding wheel in such a way that you produce a few degrees tip clearance and relief, much like grinding the bottom of a boring bar.

• Diameters of six sided cutters are easy to measure. You simply measure across any two opposing vertices and that measurement is your cutting diameter.

• They provide adequate chip clearance and leave a nice finish. You can mill deep, accurate slots with these cutters because of their relative strength and the fact that their hexagonal shape prevents galling by allowing chips to escape. Note that by default, the relief angle of a six sided single flute end mill is always thirty degrees and the rake angle is always zero degrees.

• Tip radius can be applied to these cutters with a single flute cutter grinder so that they can be used for 3D contouring.

• These cutters can also be used as small boring bars for doing lathe work.

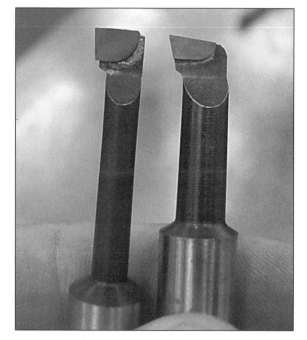

FIGURE 7–8   Small boring bars ground with a lot of clearance have less tendency to push away and make holding tight tolerances easier.

FIGURE 7–9   Relatively coarse toothed slitting saws work well for most slitting saw jobs and won't load up with chips as easily as fine toothed saws.

8. *Purchase square shanked boring bars for more versatility. (See Fig. 7-7)*

   Boring bars with flat shanks are easily mounted in lathe tool holders and in boring heads. Boring bars with round shanks are a little more difficult to set up in a lathe.

9. *Put a lot of clearance on small boring bars to hold tight tolerances. (See Fig. 7-8)*

   Boring bars come from the factory with very little side clearance as can be seen on the boring bar on the left side of the photo. Small clearance angles cause cutters to rub and push away. A tip ground with about thirty degrees side clearance significantly reduces rubbing which makes holding tight tolerances easier. Another way to reduce cutting pressure is to use a small tip radius such as .005–.010″.

10. *Use relatively coarse toothed slitting saws. (See Fig. 7-9)*

    Fine tooth slitting saws are prone to load up with chips which can create cutting problems. Coarse tooth slitting saws have better chip clearance.

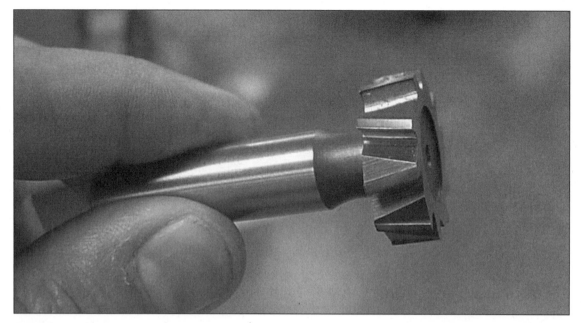

FIGURE 7–10 Stagger tooth key cutters cut with less pressure than straight toothed cutters and have less tendency to chatter.

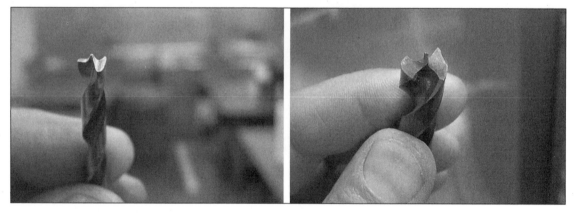

FIGURE 7–11 A sheet metal drill will cut a cleaner hole through sheet metal than a standard drill bit.

To resharpen or regrind a slitting saw or any multi-flute cutter for that matter, you'll need to rig up some sort of arrangement so that you can index the cutter to its own teeth. Be sure to grind the teeth of slitting saws with a few degrees positive rake.

11. *Use slant or stagger tooth Woodruff key cutters. (See Fig. 7-10)*

Stagger tooth key cutters are less prone to chatter and cut with less pressure than straight tooth key cutters.

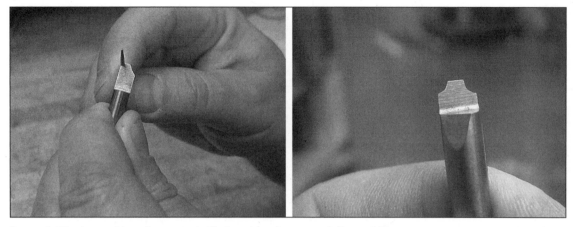

FIGURE 7–12 An outside radius cutter is filed to size using a round diamond file.

12. *Use sheet metal drills to drill sheet metal. (See Fig. 7-11)*

Sheet metal drills give you cleaner holes through sheet metal than standard drills. You can grind these drills by hand as long as you grind the drill in such a manner that the circumference of the drill is the first thing to break through the sheet metal after the center point. Also, like any cutting tool, the leading edges have to be the first things to contact the work.

13. *Use a round diamond file to make small, outside radius cutters. (See Fig. 7-12)*

Once in a while you may need an odd size outside radius cutter that you don't have. Instead of waiting to make an outside purchase you can make one.

Start by splitting the end of a round HSS or carbide blank. Then rough grind by hand or in a fixture the approximate radius into and around the bottom of the remaining half of the blank. Then use a round diamond file to finish the cutting edge radius to the size you want.

After that, it is a matter of relieving all the material behind the cutting edge by a few degrees so the cutting edge is the first and only thing to contact the work.

The most difficult part of making these cutters is the consistent clearing of the material behind the cutting edge. By taking a few gentle test cuts in some soft material you'll be able to see if the cutter is rubbing. If it is, then you can clear those areas accordingly.

14. *Use diamond files to create and shape form tools. (See Fig. 7-13)*

A set of small diamond files can be used for shaping and "tuning up" form tools. The picture shows a small form tool that was finished with diamond files and the corresponding form that was created on the shoulder of a core pin. A diamond cup wheel

FIGURE 7–13 Diamond files can be used for final shaping of form tools. The small radius form on this lathe tool was filed in with a diamond file.

FIGURE 7–14 The bottom cutting edges of an end mill are sharpened in a surface grinder with a fixture designed for the purpose. This fixture uses 5C collets to hold end mills.

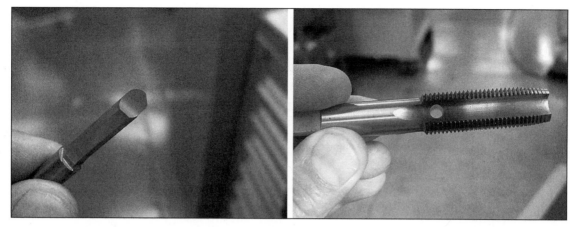

FIGURE 7–15 A carbide single flute ball nose cutter can be used to bore out broken taps. The tap on the right was bored through as an example.

mounted in a grinder can be used to rough in shapes then the files can be used for final shaping and sizing. Diamond files cut either HSS or carbide with relative ease. A set of medium and fine grit diamond files are useful tools to have.

15. *Use a sharpening fixture to sharpen end mills in the surface grinder. (See Fig. 7-14)*

The tips of end mills are usually the first things to break down. By using one of these grinding fixtures you can quickly regrind the ends of end mills in a surface grinder.

16. *Use a six sided carbide ball nose cutter for drilling out broken taps. (See Fig. 7-15)*

I have used these ball nose end mills numerous times to drill out broken taps. For my money drilling out a broken tap is less hassle than burning one out in an EDM. (Electro Discharge Machine) The easiest way to make these cutters is to use a single flute cutter grinder.

The first thing to do is to grind the hex shape on a carbide blank. The hex shape needs to be long enough so that the cutter can drill or bore all the way through the broken tap. Grind the cutter about 15% under tap drill diameter so that when you start chipping out the tap material it will come out in relatively large but easily breakable pieces.

After grinding the hex shape, the end of the cutter can be split in half and the ball nose applied. Clear the non-cutting side of the ball to reduce the tendency for the cutter to chip out. Leave the vertices of the hex sharp when making these cutters.

To use these cutters effectively you must start with a rigid setup. You wouldn't be able to drill out a broken tap in a drill press or by hand for example.

The feeding of these cutters must be done against the quill stop in tiny increments of about a thousandth of an inch at a time or less. You do that by slightly rotating the quill stop each time you peck against the broken tap. Let the cutter dwell a little on the tap material as you peck. Dwelling the cutter allows the cutter to seat and also helps soften the tap by heating it up.

Because of the uneven nature of the end of a broken tap the cutter must be feed gently into the tap in the beginning to avoid chipping the cutter. I like to run these cutters at about 3500 RPM using no coolant or cutting oil. Once you get a seat going in the end of the tap you may be able to increase feed a little bit. Use compressed air to remove chip dust as you bore. The photo on the right shows a hole bored through the side of a tap with one of these cutters.

Once you drill through the tap, it's time to start chipping out the remaining tap material with an old punch. Sometimes this process goes smoothly and other times it doesn't. A little penetrating oil may help the process. Use compressed air liberally when chipping out tap material. When you are done chipping out the broken tap then use an old tap to clean up the thread and remove any remaining tap material.

17. *Use fine grit diamond wheels for grinding carbide.*

Fine grit diamond wheels work amazingly well for grinding carbide. It's best to have a roughing wheel and a finishing wheel, but if you are only going to have one, then it should be a fine grit wheel. 150 to 180 grit wheels leave cutters with the smooth cutting edges needed to produce long tool life and smooth finishes.

Green wheels (silicon carbide) can be used for grinding carbide but they wear quickly. It is nearly impossible to grind fine forms in cutters with green wheels. Their best use is for roughing. Green wheels also work well for dressing shapes on polishing stones.

18. *Sharpen the tops of HSS lathe tools first.*

To produce the sharpest cutting edge, grind the rake angle or top side of a lathe tool first before grinding the relief and clearance angles.

19. *Avoid grinding chip breakers into the tops of lathe tools.*

Some people may disagree with this suggestion. So be it. In my experience the only thing grinding a chip breaker into a lathe tool does is substantially reduce the life of the tool.

Chip breaking has much to do with feeds and speeds. Increasing your feed and reducing your depth of cut helps curl chips. (See Fig. 7-16) Chips produced with fast feed rates help keep the workpiece cool since much of the heat of the cut comes out in the chip. I don't mind a long chip as long as it curls tightly. One way to break stubborn lathe chips is to intermittently stop and start the carriage feed.

FIGURE 7-16 A "perfect" chip that curls up tightly and breaks off in small pieces is created during this cut. It is more important, however, to make perfect parts than perfect chips.

I'm going to repeat the following relationship since I believe it provides a solid frame of reference from which to work: "From any given setting of feeds and speeds, doubling the feed rate and halving the depth of cut results in equivalent cutting times."

Another factor that influences the way your chip curls is tip radius. Smaller tip radii seem to curl chips better than larger radii. Also having the leading edge of your lathe tool set close to ninety degrees to the axis of the part helps curl chips.

20. *Use a gauge pin to check lathe tool radii. (See Fig. 7-17)*

When a drawing calls for an odd size shoulder fillet on a lathe part you may be obliged to grind the tool by hand. One quick way to check the radius as you grind the cutter is to use a gauge pin of the appropriate diameter as a reference. For example if the drawing calls for a fillet radius of .045″ then you would use a gauge pin .090″ in diameter for the check. Place the end of the pin on the tip of the tool to get an idea of how much to grind and where to touch up the radius. I like to use a fine grit hand held diamond lap for final sizing and touch up work because they cut quickly and leave a

FIGURE 7–17 A lathe tool radius is checked with the appropriate size gauge pin.

FIGURE 7–18 A template is used to check the form of this lathe part. Small templates like this can be made with brass
shim stock and scissors.

FIGURE 7–19 A trepanning tool ground at an angle provides adequate tool clearance for accessing the face of the part.

clean, sharp edge. In some cases where extreme accuracy is needed you may need to check the tool radius in an optical comparator.

21. *Make a radius gauge with shim stock. (See Fig. 7-18)*

    Occasionally you may need to file an odd size radius on a lathe part. One way to make an outside radius gauge is by sandwiching a piece of shim stock between two pieces of aluminum and drilling the appropriate size hole. After the hole is drilled you can cut the shim with scissors to create the shape of gauge you want.

22. *Use HSS or cobalt cutters to create smooth finishes.*

    HSS and cobalt single point tools such as lathe tools and fly cutter bits generally produce smoother surfaces when cutting most materials than identically ground carbide bits. Therefore, if you are trying to cut the smoothest surface possible such as when cutting molding surfaces, switch to HSS or cobalt for finishing.

23. *Grind trepanning (face grooving) tools at an angle for easier access to the face of the part. (See Fig. 7-19)*

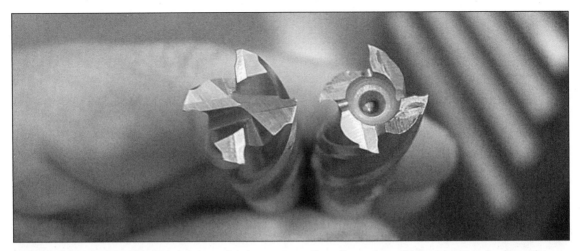

FIGURE 7–20 Since there is little tendency for the center section of a center cutting end mill to load up, they are generally a better choice for all around milling.

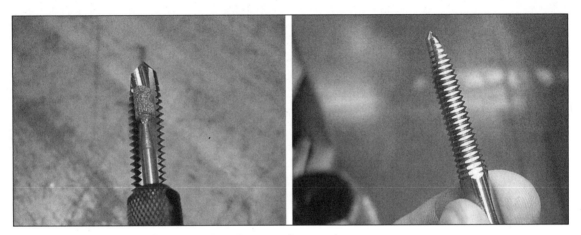

FIGURE 7–21 A tap is sharpened using a Dremel motor and mounted stone.

24. *Dull the cutting edges of drill bits and other tools that are going to be used to cut brass.*

Brass, like no other material I've come across, has a tendency to grab and suck in positive rake cutting tools. The tendency for brass to grab cutters can be outright dangerous at times and may suddenly yank the cutter or workpiece out of position.

To keep cutting tools from grabbing while cutting brass you should break the cutting edges with a stone or grind a little negative rake on the cutting edges.

25. *Purchase and use center cutting end mills. (See Fig. 7-20)*

I find myself choosing center cutting end mills when given a choice. Non-center cutting end mills work for many jobs but may unexpectedly load up and create problems.

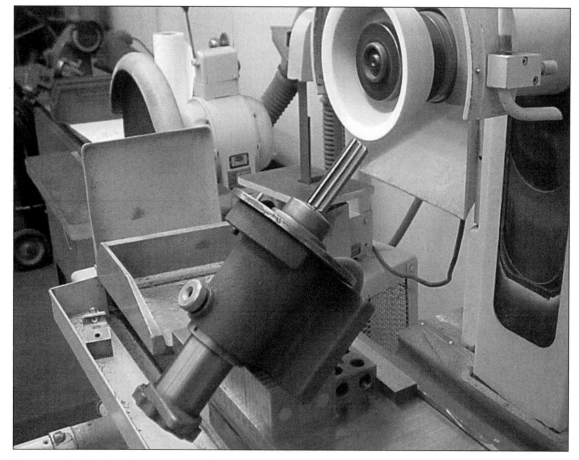

FIGURE 7–22 The tips of reamers are the first things to wear out and can be resharpened in a surface grinder.

26. *Regrind taps with a Dremel® Tool. (See Fig. 7-21)*

If you want to get more life out of dull taps you can regrind them by hand using a Dremel motor with a mounted stone. The trick is to grind a little positive rake into the cutting edges or teeth of the tap. To do that, you need to roughly match the diameter of the stone to the size of the flute. Try to grind each flute equally. Grind with the spindle rotating in a direction that keeps the stone from walking over the edges of the teeth. Once you get the hang of sharpening taps this way, you'll notice that they cut well with little pressure. Note that taps ground with negative rake require much more torque to wind in than those ground with positive rake.

Another trick to reduce the torque needed to wind in a tap is to grind flats across the non-cutting diameters of the taps as shown in the right photo.

27. *Regrind reamer tips. (See Fig. 7-22)*

It would be difficult for machinists to live without reamers. They are time savers. You'd be hard pressed to find another method that works so quickly to accurately size holes.

The tips of reamers do most of the cutting and are the first things to wear. It is relatively easy to grind reamer tips in a surface grinder, especially with larger diameter reamers. You can either dress an angle on a grinding wheel or tilt the reamer as shown in the photo. A forty-five degree lead-in angle using plenty of cutting edge relief works well.

28. *Use your fingers to check a cutter's edge condition.*

Gently pull your finger across a cutting edge at right angles to the edge. Sharp cutters will try to hang onto the surface texture of your fingers. Your finger will slide over dull edges with little resistance.

29. *In aluminum, use a two flute end mill for roughing and a four flute end mill for finishing.*

This suggestion may raise some eyebrows. So be it.

Machinists learn at an early age that two flute end mills are used to cut aluminum and other soft materials.

The main reason for using two flute end mills is to avoid chip packing. With the harder T6 aluminum alloys commonly used today, I find that chip build up is rarely a problem especially with larger diameter end mills.

I can think of a lot of reasons why it may be preferable to use a four flute end mill even when cutting aluminum.

- Two flute end mills are not as rigid as equivalent diameter four flute end mills because of their smaller cross sectional area.

- Given equivalent feeds and speeds, a two flute end mill will have twice the chip load as a four flute end mill. Because of the increased chip load under similar cutting conditions, cutting pressures are greater. The only way to reduce chip load is to reduce feed rate or increase spindle RPM each of which may have negative consequences.

- Given equivalent feeds and speeds your finish won't be as good with a two flute end mill.

- If a two flute end mill develops a flaw or already has a flaw in one of the flutes, there is only one other flute working to remove the flaw. With a four flute end mill there will be three other flutes working to remove the flaw left by the bad flute.

- Four flute end mills have better stability when used for hole sizing and counter boring. Four opposing flutes plunging straight down are more self centering and have less tendency to dance around than two flutes.

Nevertheless, if you are cutting in an area where chips have trouble escaping like a deep slot or pocket or when taking a full diameter width cut then two flute end mills are a better choice to help alleviate chip packing. For the best of both worlds in aluminum and other soft materials, use two flute end mills for roughing and four flute end mills for finishing.

30. *Use high helix carbide end mills to produce smooth surfaces. (See Fig. 7-23)*

Although a little pricey, high helix carbide end mills work well for cutting hard and difficult to machine materials such as stainless steel. They also produce superior surface finishes. From an economy standpoint it is best to use some other "thrasher" end mill to rough with. Save your high helix end mills for finishing.

31. *Shorten the tips of center drills. (See Fig. 7-24)*

How many times have you inadvertently broken the tip off a center drill? It doesn't take much does it? A little bit of run-out is generally all that it takes to break a factory ground tip. If you grind the tip of a center drill to about half the factory ground length, it will be much harder to break.

FIGURE 7–23 High helix end mills are rigid and produce smooth milled surfaces.

32. *Be aware that HSS will cut with less pressure than carbide.*

This is one of the great mysteries of the universe. Given identically ground cutters of HSS and carbide under identical cutting conditions, carbide will cut with more pressure for some reason.

33. *Use a left hand lathe tool to make a right hand lathe tool. (See Fig.7-25)*

This is a trick I learned not too long ago.

The picture shows a left hand brazed carbide lathe tool ground to be used as a right hand tool. Cutters ground like this work

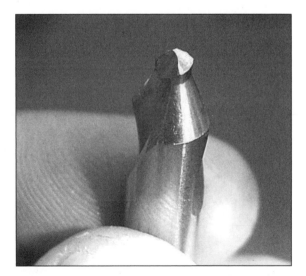

FIGURE 7–24 The tips of center drills have less tendency to break if they are ground to about half the factory ground length.

great, are more resistant to breaking and are easier to sharpen than conventionally ground lathe tools.

34. *Is it OK to grind* HSS *with a diamond wheel?*

I've been finishing HSS cutters with diamond wheels for years and have never had a problem. I like to use diamond wheels for finishing HSS because of the smooth, sharp edges they produce. Smooth sharp edges promote cutter longevity and efficient machining.

Some people claim that grinding HSS with diamond wheels loads up the wheels and renders them useless. I've never seen evidence of it. Just to be safe, avoid using diamond grinding wheels to rough off HSS.

FIGURE 7–25 A left hand brazed carbide lathe tool is ground as a right hand tool. Tools ground like this resist chipping and are easy to sharpen.

35. *Use "middle of the road" grade carbides for the majority of your machining.*

This is a complex subject that I prefer to keep simple. Under laboratory and high production conditions you could probably increase material removal rates and tool life by experimenting with different grade carbides, coatings and chip breakers.

For "run of the mill" low production and prototype machining, good quality straight tungsten carbide works well. "Micrograin" carbides seem to hold an edge a little better when machining hard materials. Hard grade carbides such as C8 are very brittle and chip easily. Since I am an advocate of high feed rates for removing metal quickly, I prefer the softer but tougher grade carbides such as C2 and C5 that don't chip so easily.

Assuming reasonable feeds and speeds, the rigidity of your setup, part and cutter and your ability to avoid "crashes" will have more to do with how long your cutters last than just about anything else. Another important factor regarding cutter life is chip packing. Chips that can't easily escape are destined to be recut which can significantly reduce cutter life.

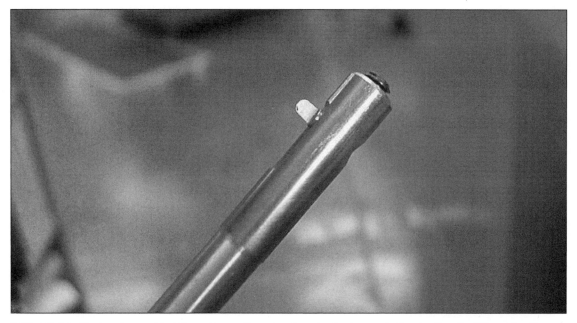

FIGURE 7–26 Ejector pins are ideal for making cutter extensions, boring bars and other tooling. To drill and tap through the side of an ejector pin you must first grind through the case hardened surface.

36. *Use a piece of ejector pin to make boring bars and other tools. (See Fig. 7-26)*

The rigidity, accuracy, and hard surface of ejector pins make them ideal for constructing boring bars, and cutter extensions. To drill through the side of an ejector pin with a HSS drill bit, you must first grind through the case hardened surface of the pin.

37. *Make special cutters from standard end mills. (See Fig. 7-27)*

One useful tool you can make from an end mill is a 45° cutter.

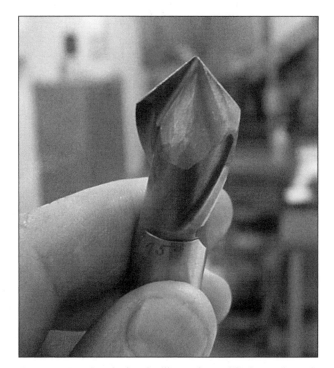

FIGURE 7–27 Standard end mills can be modified to make various cutters. The 45° cutter was made from a flat bottom two flute end mill.

Threads and Things

What a great invention. Threaded parts have been around for centuries and will likely be with us indefinitely. Threaded parts work well, have numerous uses and are relatively easy to produce.

Historically, standardization has been one of the main stumbling blocks of threaded parts and to some extent is still an issue. In 1918 a number of professional groups in the U.S. got together and settled on a standard that may still be used called the American National Standard. The geometry of the American National thread is easy to work with in that the root and crest measurements are specified as one-eighth of the pitch.

Unified Thread

Over time it became obvious that the American National thread had some drawbacks. Namely the tolerances are too tight for easy manufacturing, the threads were not interchangeable with those of other Allied countries until 1948 and there is no provision for tool wear which produces a radius at the root of the thread.

So a group of Oracles from the Allied countries got together and came up with an improved standard called the Unified thread. The Unified thread, which was adopted in December of 1948, is a loosened up, reverse engineered version of the American National thread with a provision for tool wear so that the root and crest of a Unified thread may either have a flat or a flat with radii or a full radius. Parts made to either standard are interchangeable.

Threads made within the tolerances of the American National thread are also within the tolerances of the Unified thread. The reverse is not true however.

Bear in mind that although the old American National thread has been outdated, it has not been outlawed. Any threads called out on a drawing that do not include the letters "U", "A" or "B" may be considered to be made to the old American National standard. For example, if your print unerringly calls for a "1/2–13 NC-2" thread instead of a "1/2–13 UNC-2A", in theory you're obliged to work to the tolerances of the older standard.

One of the drawbacks of the Unified thread aside from not being interchangeable with metric threads is that the geometric relationships of the thread are somewhat more complicated than the American National thread.

UNR Thread — Unified National Round (Round Root)

Remember that with the Unified thread standard adopted in 1948, root and crest radii are optional. The UNR thread, however, mandates a radius at the root of the thread which makes sense from a fatigue standpoint. Sharp corners on any stressed part are always more prone to failure.

Pitch Diameter

Thread geometry can be mind boggling when looked at in depth. If you look at the schematic of mating threads in *Machinery's Handbook* with all the nomenclature and relationships, you may begin to wonder how anything gets done.

Since machinists are the ones called upon to quickly and unerringly make this stuff, we have to try to make some sense out of it.

Pitch diameters are one of the most important measurement in terms of allowing male and female threaded parts to go together. The pitch diameter is the measurement between the two theoretical lines on each side of a bolt that would go through the exact middle of perfectly sharp threads. The basic pitch diameter is the diameter from which all tolerances begin.

The thread pitch and sixty degree angle are also important, but for the purpose of this discussion I'm going to assume that any taps, machine tools, cutters or other gizmos you use to cut threads will give you correct pitches and angles.

The side to side and up and down slop you feel between a nut and a screw is technically called "allowance." This allowance or slop is determined by the difference in the pitch diameters between the nut and the screw assuming the roots and crests of the opposing members have adequate clearance.

Different classes of threads allow for different amounts of slop. Standard shop practice would be to work to a class 2A (external) or class 2B (internal) thread unless otherwise stated on the drawing. Class 1A and 1B threads provide the most allowance between mating threads. Some of the closer tolerance stuff is produced to Class 3A and 3B standards which at one extreme allow for mating threads with zero allowance.

Pitch diameters can be measured in a number of ways. One easy way to check the pitch diameters of both internal and external threads is with "Go, No-Go" gauges. These gauges also check thread form to a certain extent. They are not fool-proof however. You cannot make the assumption that just because your thread passes a Go, No-Go gauge test that the thread is in tolerance. The thread must be within the limits of form of the standard you are working to. Probably the easiest way to check form is with an optical comparator.

One might ask how the form of an internal thread can be checked. One way is by taking a resin casting of the thread then checking the casting in a comparator.

One of the disadvantages of Go, No-Go gauges is that they are relatively expensive and odds are most shops won't have a complete set available for machinists to use.

Another way to check pitch diameters is with thread wires. This is probably the cheapest and most commonly used method for accurately checking pitch diameters. One quick way to use thread wires is to take a reading over an existing bolt. That way you won't have to calculate anything. It probably is safer in the long run to do the calculations. Thread wires can only be used to check external threads and although they can give you a precise reading, they require some dexterity to use. You darn near have to be a magician to hold the thread wires in position and at the same time take a micrometer reading. You can buy little gizmos to hold thread wires in place on your mikes to make the measuring process easier.

Root and Crest Geometry

Now let's take a look at the root and crest geometry of the Unified thread. This is where things can get a little murky.

The way I see it, the most important thing regarding root and crest geometry is that the opposing male and female roots and crests do not interfere with each other when pitch diameters are within tolerance. Otherwise the nut and bolt wouldn't go together.

The main things to look for on both internal and external threads are crests that are too sharp and roots that are too blunt. If either condition is excessive, unacceptable interference may result even when pitch diameters are correct.

In order not to produce too sharp a crest on an external thread, I like to use what I call the "one percent" rule. It's my own rule so you probably won't find it anywhere else. I like to reduce the nominal outside diameter by at least one percent before I cut a thread. For example, if I'm getting ready to cut a 1/2–13 UNC -2A thread, I would turn the OD to .5 – 1% of .5 = .495 or for a 3/4–10 thread I would turn the OD down to at least .75 – 1% of .75 = .7425. On the other hand, you can use *Machinery's Handbook* for this information if you prefer.

I'm not sure how the "one percent" rule holds up under all classes and sizes of thread in terms of producing a major diameter that is within Unified thread tolerances but for the run of the mill class 2A UNC and UNF stuff it works.

Now, let's turn our focus to root geometry. One of the realities of thread cutting is that any pointed cutting tool will wear to some degree creating a radius at the tip of the tool or increasing or distorting whatever radius or shape was already there.

One might wonder if perfectly sharp thread roots are acceptable. Or to put it another way, are perfectly sharp thread roots within the limits of form of the Unified thread? The answer, is no. To be technically correct, the root has to be either flat or have a flat with radii or a full radius each of which have limits. With that said, I've never seen a thread rejected because the root geometry was too sharp, but it could happen.

Starting Point

Here's the deal with machinists. We have to produce. We haven't got all day to debate thread theory. To start making chips we have to pick a starting point and go. To keep things as simple as possible in terms of creating a thread that is within Unified thread tolerances there are a couple things you can do. You can buy inserts that already have correct root and crest geometry for a particular size thread or you can start with a sharp pointed tool and create your own root geometry.

By lapping a radius on the tip of a sharp pointed 60° tool of not less than .108 times your thread pitch distance you'll have a tool that will produce root geometry that is within UN or UNR thread tolerances when pitch diameters are correct. A tip with that radius is on the sharp side of correct root geometry which allows for additional tip wear as the tool is used.

Let's make a thread on paper by using what I'm going to dub the "quick start" method.

To produce a 1/2–13 UNC -2A external thread using this method, you start by reducing the nominal outside diameter by one percent. That would give you a starting or major diameter of .495″. Next you lap a radius on the end of your threading tool equal to or slightly greater than .108 times the thread pitch distance which at thirteen threads per inch is .077″. That would give you a tip diameter of about .016″. Now you start cutting the thread until your pitch diameter falls into the median range of the pitch diameter limits given in *Machinery's Handbook* for a class 2A thread. In this case the median is .446″. The limits are .4485″ and .4435″.

That's it. That thread should be well within class 2A UN or UNR thread tolerances in all aspects. How's that for simple?

Now let's take a look at how thread root geometry may fall out of tolerance as the tool bit wears. Assuming you maintain a correct pitch diameter, one way the root geometry may fall out of tolerance is by exceeding the maximum allowable root diameter. For a 1/2–13 UNC -2A, the maximum minor diameter is given in the tables in *Machinery's Handbook* as .4069″.

Basic Pitch Diameter

Before I go on, I would like to provide a formula that quickly yields the basic pitch diameter of any Unified screw or bolt based on the nominal size of that screw or bolt.

Although this information can be found in *Machinery's Handbook* I prefer using formulas when I can to avoid the task of searching through tables.

The basic pitch diameter is useful as the starting point from which all tolerances begin. It is also the minimum pitch diameter for all classes of internal threads and is the maximum pitch diameter for a class 3 external thread.

In other words, the basic pitch diameter is the high extreme for an external thread and the low extreme for an internal thread. Parts cut to those extremes, in theory, would go together with no allowance.

The formula is as follows:

Basic Pitch Diameter = Nominal Screw or Bolt Size – (Thread Pitch Distance X .6495)

Hence the basic pitch diameter for a $1/2$–13 thread would be:
$$.500 - (1/13 \text{ X } .6495) = .4500$$

For a $1/2$–20 thread the basic pitch diameter would be:
$$.500 - (1/20 \text{ X } .6495) = .4675$$

Knowing the basic pitch diameter for a given thread gives you a solid frame of reference from which to work. Unless you are cutting class 3 threads at the upper limit which is the basic pitch diameter, any external threads you cut should end up somewhat under the basic pitch diameter and any internal threads you cut should end up somewhat over the basic pitch diameter.

Commercially available taps are ground to basic pitch diameter or a little over depending on the "H Limit" of the tap. Nominal screw diameters for number screws can be determined by the following formula:

Nominal Screw Diameter = (.013 X The Screw Number) + .060

Hence the nominal screw diameter for a #10 screw would be:

$$(.013 \text{ X } 10) + .060 = .190''$$

Inspection

As you can see threads can be paradoxically simple and complicated at the same time. Ninety nine percent of the time nobody questions them. Parts inspectors may check them with Go, No-Go gauges and if they pass that is usually the end of it.

Nevertheless, there is reason to be careful. I'd like to relate a somewhat sad but true story to illustrate my point.

One time I was looking for a shop to do some machining for a project I was working on. In one industrial complex, I walked into a small machine shop hoping to get some bids on some parts. As I entered the shop, I saw an older gentleman fussing over some parts that were all lined up on a table. There were about a hundred parts that looked somewhat like small engine carburetor bodies. They had all sorts of machined features in and on them including one relatively large boss with external threads. The parts had already been processed and had a heavy brown coating on them.

After introducing myself, the gentleman started telling me about the parts. He told me how well the parts had run on his CNC machining center and that he was supposed to have gotten ninety dollars a piece for them. I asked him what the problem was and he pointed to the external thread on one of the bosses. Upon closer examination I could see that the crests of all the threads were sharp and slightly bent to one side.

I asked him if they could be fixed and he told me not with the processing that had already been done to them. I asked him if he had used "go, no-go" gauges to check the threads and he said yes but an inspector rejected the threads because the shape of the threads were out of tolerance. Then I asked him what he was going to do and he said he didn't know. I suggested he try contacting the engineer in charge of the project to see if the parts could still be used. With that I excused myself and wished him luck.

A few months later I noticed he was no longer in business. So we learn that if threads are going to be inspected, among other things they have to have correct pitch diameters, correct major and minor diameters, correct root and crest geometries, and correct angles. (All of which the "quick start" method previously discussed will give you.) Also, a first article inspection before processing may save a lot of grief later on.

1. *What are tap "H" limits?*

   To provide some kind of adjustment for tapped holes somebody came up with the concept of H limits. An H1 tap is ground to a pitch diameter of from basic pitch diameter to plus .0005″. An H2 tap is ground from plus .0005″ to plus .001″ and so on. In other words, the higher the "H" limit the sloppier the tapped hole. A little slop is often a good thing especially when parts distort during heat treat.

2. *Use a knurled tap and die holder for threading in a lathe. (See Fig. 8-1)*

   This is a common, easily made tool you'll often see used in a machine shop for holding die buttons to thread small parts. The body of the die holder slides on a shaft held in a tailstock chuck. The shaft keeps the die aligned perpendicular and concentric with the work yet lets the die holder slide and turn freely.

   The operator must push the die onto the work as the lathe spins while holding counter-torque against the torque generated by the cut. Once the part has been threaded up to a shoulder or dimension the operator can release the die holder which will then turn freely. The threading die and holder can be unscrewed from the part by putting the spindle in reverse while holding the die holder from turning. Remember nominal outside diameters should be reduced at least one percent for best results.

FIGURE 8–1 In the left photo, a die button mounted in a guiding jig is used to finish an external thread. In the right photo, a hole is being tapped with a guiding jig.

FIGURE 8–2 Adjustable die buttons allow for roughing and finishing a thread to the correct pitch diameter.

Tapping can be done in a lathe with a similar arrangement. Some provision must be made for holding taps instead of die buttons. One way is to use a common drill chuck mounted on a hollow sleeve.

3. *Purchase high quality, adjustable die buttons. (See Fig. 8-2)*

   Adjustable die buttons allow you to rough and finish a thread and make pitch diameter adjustments.

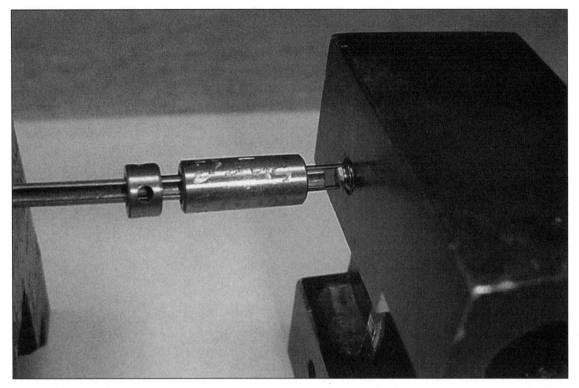

FIGURE 8–3   Tap removers seldom accomplish the task of removing a broken tap.

4. *Use two flute taps for maximum strength.*

They have a greater cross sectional area than four flute taps and are therefore stronger.

5. *Forget about using tap extractors to remove broken taps. (See Fig. 8-3)*

In the thirty years I've been working as a machinist I've never seen a tap extractor accomplish the task of removing a broken tap. I've seen many fail to do so however. The "ears" of the tap extractor invariably break before enough torque can be applied to remove the broken tap. The photo shows a tap extractor in the process of being installed over a broken tap. (See Chapter 7, item #16 to find out how to remove broken taps.)

6. *Feed your single point threading tool in at an angle to reduce right angle pressure.*

Cutting pressure can be a significant issue when cutting threads with a single point tool. Even by supporting the end of the part with a live center the middle portion of the shaft may push away from the threading tool if the part is relatively small in diameter. One way to reduce the tendency for the part to push away is to feed the tool in at a 29.5° angle so that some of the pressure of the cut is directed toward the spindle.

7. *Use a geometric die head in a lathe for accurately cutting external threads on small parts. (See Fig. 8-4)*

Geometric die heads are great tools for accurately and consistently threading small parts in semi-production quantities. Once a setup is made, these tools are easy to use and consistently produce quality threads. The die heads have provisions for roughing, finishing, precisely controlling pitch diameter and thread length.

8. *Make an external thread gauge with a tap. (See Fig. 8-5)*

FIGURE 8–4 Geometric die heads can consistently produce accurate, high quality threads.

As noted earlier, I've seen people check external threads with common hardware store nuts and be perfectly happy with the results. Using common nuts is not a great way to check external threads because the nuts may have threads that are significantly over basic pitch diameter which would allow the external thread to also be over basic pitch diameter. External threads should never be over basic pitch diameter unless the thread is special case and made for an interference fit.

An accurate thread gauge can be made by buying a tap at the H1 limit and using that tap to make a gauge. Remember H1 taps are ground at from basic pitch diameter to plus .0005″. Any external thread having minimal slop with such a gauge would likely be within class three tolerances.

9. *Cut single point threads away from the shoulder of a lathe part to be safe. (See Fig. 8-6)*

Sometimes it can be difficult using a conventional lathe to consistently stop the threading tool in the heart of the undercut. One way to remedy that difficulty is by feeding from the undercut out towards the tailstock. Turn the threading tool upside down and cut the thread with the spindle running in reverse. Hardinge® type tool room lathes have an automatic "kick out" feature that eliminates the problem of stopping the tool in the undercut.

FIGURE 8-5   A decent external thread gauge can be made by carefully tapping a piece of material.

FIGURE 8-6   Single point threading with the cutter inverted and the spindle running in reverse allows feeding the tool from the undercut towards the tailstock.

FIGURE 8–7   Holes can be hand tapped by loosening the spindle collet.

10. *Avoid disengaging the half nut of an old engine lathe when single point threading.*

Some conventional lathes have half nuts that are sticky and difficult to engage. The half nut is what you engage when you throw the lever to begin single point threading. Instead of risking ruining your thread with a bad engagement, keep the half nut engaged at all times during the course of cutting the thread. When you are done with a pass, stop the spindle, clear the tool, then reverse the spindle to go back to the beginning of the thread. If the thread is next to a shoulder, feed from the shoulder toward the tailstock with the tool upside down as described above.

11. *Switch to HSS or cobalt when single point threading if you are having difficulty getting a good finish with carbide, or if you are having difficulty with your carbide threading tips chipping.*

12. *Open up minor diameters for easier tapping.*

When tapping tough material, use a tap drill one or two number sizes larger than a 75% thread chart would call for. In tough material, an opened up minor diameter substantially reduces the torque needed to wind in a tap. The only time I don't go bigger with a tap drill is when a part is thin or when the material is soft.

13. *Use a straight shank drill chuck in a loose collet to manually tap holes. (See Fig. 8-7)*

This method works well for tapping holes in tough material because you can control the torque you put on a tap. The loose collet provides support and alignment for the shank of the chuck yet lets it turn freely so that you don't have to rotate the machine spindle. Taps can take more torque before breaking if no side load is put on them.

FIGURE 8-8 A tapping block is used to line up a tap for hand tapping.

14. *Use a tapping block to guide taps when hand tapping. (See Fig. 8-8)*

A tapping block prevents taps from getting started crooked.

15. *What is a fifty percent thread?*

In approximate terms when the minor diameter of an internal thread is drilled out so that half the thread height is left, that is a fifty percent thread. More precisely, an internal thread cut to basic pitch diameter then drilled out or truncated to that diameter is a fifty percent thread.

According to one source, for a given material, a fifty percent internal thread tapped to a depth of one and a half times bolt diameter has greater holding power than the force needed to pull apart a bolt made of the same material.

FIGURE 8–9   Stick wax lubricant adheres to a taps even under a coolant stream.

16. *Use stick wax to lubricate taps in CNC machines. (See Fig. 8-9)*

Coolants used in CNC machines usually don't have great thread cutting properties. Common band saw blade stick wax applied to taps used in CNC machines will adhere to taps and resist being washed away in a coolant stream. The stick wax reduces friction and tap failure. Another way to reduce tap failure is to program the machine to stop before a tapping cycle begins so an operator can apply cutting oil to the tap. The coolant stream should be turned off when using cutting oil so that it doesn't get washed away.

17. *Rough in a thread with a single point tool then finish it with a die button. (See Fig. 8-10)*

This is the method I prefer to produce clean accurate threads on short run parts.

Die buttons work best when cutting small amounts of material. Rarely can you just force a die button set to finish size over a shaft and expect a decent thread. On larger

FIGURE 8–10 Threads can be machined in a variety of ways. This thread was roughed in with a single point tool and finished with a die button.

threads, die buttons can generate high and even dangerous cutting forces. One way you can reduce those cutting forces and at the same time produce a clean, parallel thread is to rough in the thread with a single point tool then finish it with a die button.

18. *Roll threads for maximum strength.*

   External thread rolling is a specialty and is not something you would normally do in a typical machine shop. Thread rolling produces a strong clean thread since the grain of the material is forced to flow with the contour of the thread and is not interrupted like a cut thread.

19. *Roll form internal threads with chipless taps in ductile material. (See Fig. 8-11)*

   In one shop I worked, chipless taps were used with tapping heads almost exclusively for threading aluminum. It's not a bad way to go since the taps are strong, the resultant threads are clean and strong, and no chips are produced.

FIGURE 8–11 Clutch type tapping heads work well for production tapping.

20. *Make a simple tap extension. (See Fig. 8-12)*

    Bore the end of a shaft for a slip fit with the shank of a tap then mill a slot beyond the bored hole that engages the drive square.

21. *When tapping tough material use different styles and brands of tap in the same hole.*

    Tapping tough materials requires patience. Switching from one brand or style of tap to another as you hand tap deeper into a hole will make the tapping easier. Intermittently switching from a spiral point tap to a bottom tap works well.

    Hangsterfer's Hardcut® Series cutting and tapping oils are especially well-suited for tapping tough materials.

22. *When using die buttons or thread chasers use thick, black, smelly (sulfurized) cutting oil for best results.*

23. *Avoid power tapping with plug taps and hand taps.*

They will bind if you do because the chips lodge between the flutes. It is better to use a spiral pointed tap as shown in Fig. 8-12 for power tapping because chips get pushed ahead and don't cause binding. Spiral point taps are also known as "GUN ® tap."

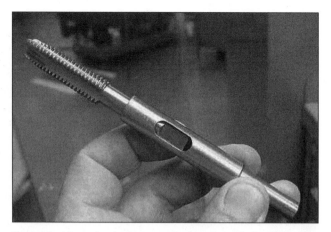

FIGURE 8–12 A simple tap extension can be made by drilling the appropriate size hole in the end of a shaft and milling a drive slot.

24. *Use a sixty degree triangular file to clean up damaged threads. (See Fig. 8-13)*

FIGURE 8–13 A damaged thread is cleaned up with a 60° triangular file.

GUN® is a federally registered trademark of Greenfield Industries, Inc. to describe taps manufactured and sold by Greenfield Industries, Inc.

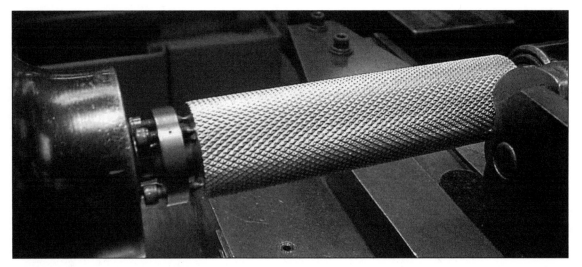

FIGURE 8–14 A clean knurl is produced.

25. *Interpret thread call outs correctly.*

   In theory, a designer or draftsman should call out tap drill depth. Many draftsmen only provide the last half of the call out and leave the drill depth up to the machinist. Interpret 1/4–20 ⟁ .6 to mean that the designer doesn't care how deep the tap drill goes within reason and that he wants at least .6″ of full thread.

26. *What does the "6H" refer to in a metric thread call out such as: M6 X 1 – 6H?*

   It refers to the class of thread which can be referenced in *Machinery's Handbook*.

A Note About Knurling

As many of you know, knurled parts are used in wide variety of products. Knurled surfaces are commonly seen on cylindrical parts such as handles, knobs and rollers. Knurling is a process that creates a pleasing, functional, geometric pattern of some sort on the surface of a part.

   Knurling is accomplished by using a tool that uses two freely rotating rollers that deform or push material rather than cut it. Creating a crisp, clean knurl in a lathe is easy if you employ a few of the following suggestions. (See Fig. 8-14)

27. *Understand the relationship between a knurling tool and the diameter of the part to be knurled.*

   This is where some people get into trouble. They try to start a knurl over an arbitrary diameter and find that the knurl doesn't track in the previously laid grooves.

The relationship is as follows:

A proper diameter to knurl is any diameter that is a multiple of the spacing or distance between the teeth of a knurling tool divided by PI. (3.14159)

The relationship is the same whether the knurl is a diamond or a straight knurl. However, the spacing of the teeth of a diamond knurl must be measured along the axis of the part or roller for the relationship to hold true.

Let's do an example:

Suppose you want to impress a diamond knurl on a one inch diameter shaft. Suppose also that the spacing or distance between each tooth of the knurling tool measured along the axis of the roller is approximately .060″. You can measure the spacing with calipers. The measurement is not extremely critical in that ultimately your final diameter will be determined by trial and error. However, the measurement will give you a decent starting point.

According to the above relationship, if we divide .060 by PI (3.14159) we get .019. Accordingly, any multiple of .019 should give you a diameter that would give you a perfect knurl. For example: .019 times an arbitrary number such as 40 equals .760. In theory then, if you turned a shaft to .760″ you would be able to create a perfect knurl.

However, since we want to knurl a shaft that is approximately one inch in diameter we have to find a multiple of .019 that gets us close to one inch. After a little trial and error and playing around on a calculator we find that .019 times 52 equals .988 which is close to one inch. .988 then would be a good theoretical starting point.

In practice though and from experience the chances of getting a perfect knurl on the .988 shaft diameter are not great. The error happens as a result of an imperfect measurement made between the teeth of the knurling tool which is no big deal anyhow because ultimately you are going to sneak up on a usable diameter. Begin by machining the part about .010″ larger than the calculated diameter. Let's proceed.

28. *Lay a short test knurl by hand. (See Fig. 8-15)*

Once you've machined a diameter that is a few thousandths larger than the calculated diameter you can begin testing.

Mount the knurling tool approximately perpendicular to and on center to the workpiece. It is not extremely critical that the tool be either exactly perpendicular or exactly on center to work properly.

Start a test knurl by lightly pressing the rollers of the knurling tool against the work. I like to start knurling close to the headstock since that is where the part is most rigid. Rotate the lathe spindle slowly by hand as you watch the pattern being impressed on the work.

As one turn of the spindle is completed the pattern will start to repeat itself on the workpiece.

FIGURE 8–15 A test knurl is laid by hand to see if the diameter produces a knurl that tracks in previously laid grooves.

If the grooves impressed on the part near the beginning of the second rotation don't line up with the grooves impressed on the part from the first rotation, that's an indication that you'll have to make an adjustment to the diameter of the work.

Machine a couple of thousandths off the diameter of the shaft and do another test. A little over one turn is sufficient to see if you have a winner. Once you find a diameter where the grooves fall on top of each other, record the diameter and proceed. The rest is easy.

Put some pressure on the knurling tool and start feeding the tool slowly along the length of the shaft. Feeding slowly on the first pass helps maintain proper tracking of the rollers in the previously laid grooves.

29. *Use feeds, speeds and pressures you feel comfortable with.*

These parameters are not overly critical. Start with slow to moderate settings then increase them as you see fit. A lot depends on the rigidity of your part and setup. In areas where the knurl comes in shallow you can dwell and concentrate the tool in that area to help balance the groove depth.

30. *Use air and lightweight lubricating oil for knurling.*

Since knurling is not a cutting operation, it's best to use a lubricant that aids in slipping not cutting. You want the teeth of the knurling tool to slip back into previously laid grooves.

I like to use lightweight oil such as WD-40® so the oil doesn't retain the inevitable flakes of material that are created during the knurling process.

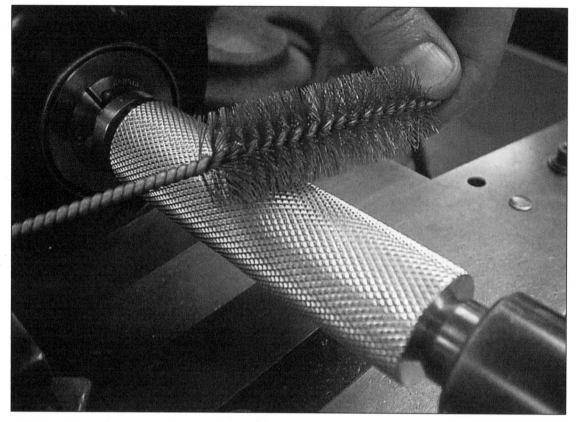

FIGURE 8–16 A wire brush is used to remove flakes of material that have pressed into the knurl.

Blowing a hard stream of air over the rollers and workpiece during the operation helps prevent flakes from getting pressed into the knurled surface.

31. *Run a smooth file over the top of the knurl to remove sharp points.*

A sharp knurl can be uncomfortable to handle. File or sand the knurl to adjust for the feel and diameter you want.

32. *Finish the knurl using a wire brush to remove burrs and flakes. (See Fig. 8-16)*

Help for
Engineers

chapter

From a Machinist's Perspective

I find it odd that thousands of management hours are spent discussing organizational and scheduling issues and very little time is spent by manufacturing personnel discussing technical issues. As a result, we see people in engineering and manufacturing making the same mistakes over and over.

In this chapter I'm going to provide suggestions to designers and engineers that seldom get feedback from shop personnel. My goal is to help designers improve the little things that add up to make manufacturing easier and more productive for all involved.

1. *Avoid verbal instructions.*

    Unless you are immediately available to answer questions (seldom the case) it is best to avoid verbal instructions. As much as I hate drawings scribbled on napkins, I'd rather have a napkin drawing than nothing at all. When you give a machinist or anybody else for that matter, verbal instructions and the job doesn't turn out right, don't blame anybody but yourself. There is a high probability verbal instructions will be misinterpreted.

2. *Avoid napkin drawings and other carelessly scribbled drawings.*

More often than not they are either wrong or incomplete. The fact of the matter is most hard copy drawings have errors or are incomplete. My guess is that at least half of all drawings circulating in manufacturing don't provide enough information to complete a job. As simple as it would appear, it is difficult to dimension a print from scratch and provide all the information needed. I wouldn't be surprised if some of the project drawings I provided at the back of the book are incomplete. One suggestion that might help alleviate the "missing" dimension problem would be to provide the shop with a CAD file along with a hard copy.

3. *Update drawings.*

When you bring a revised drawing to the shop for work in progress, highlight the dimensions that have been changed.

4. *Use decimal places wisely.*

Decimal places are a good way to convey to a machinist the tolerance or importance of a dimension. It is aggravating to find out, after the fact, that a drilled hole would have worked fine in a part that you meticulously bored or reamed because a four place decimal was called out on the drawing.

5. *Avoid using welds as a means of attachment in low quantity assemblies.*

Welding almost always distorts parts. From what I have seen, most designers don't know that. To avoid the inevitable frustration of trying to straighten and/or re-machine welded parts, designers should know that from a manufacturing standpoint for small precision assemblies, it is often easier and more accurate to use screws and dowels instead of welds as a means of attachment. If you do use welds in your designs then make it clear on your drawings that final sizing is to take place after welding.

6. *Avoid dimensioning holes and features from more than one origin.*

From a design standpoint dimensioning from more than one origin may not matter. From a manufacturing standpoint it is almost always easier to work from one origin.

7. *Design using coarse threads in aluminum.*

Coarse threads in aluminum are stronger and last longer than fine threads, especially when the threads are used repeatedly.

8. *Avoid using 6-32 threads in tough material whenever possible.*

   6-32 taps have a small cross sectional area relative to the size of the thread they cut and are therefore more prone to breaking than other size taps. It is easier to tap a 4-40 or 8-32 thread than a 6-32 thread in tough material.

9. *Avoid designing with irregular spacings between holes and features when possible.*

   When working with hardware it is nice to be able to use a scale or caliper to quickly verify feature spacing.

10. *Avoid scattering dimensions all over a drawing.*

    I prefer working from a drawing that has as many dimensions as practical in one view.

11. *Don't bury important information in the general notes.*

    Even though we should, machinists don't always read the general notes on a drawing. If there is an important bit of information that must be read for the proper machining or processing of a part, then it is best to put a flag in the body of the drawing to call attention to the note.

12. *Call out screw clearance geometries.*

    Since not all screw clearance holes are created equal, I prefer that a designer state on a drawing what size hole and counter bore he wants for a particular screw.

    For example,

use:	Ø .281 Thru, ⌴ Ø .406 ▽ .28″
or:	Ø .265 Thru, ⌴ Ø .391 ▽ .28″
instead of:	Drill and C'bore for 1/4–20 SHCS

13. *Call out on a drawing whether you want holes drilled or reamed.*

    This is not always necessary but a call out to drill or ream a hole removes any doubt as to what you want. For example a designer could make the following call out on a drawing to aid the machinist: Ø .5000 Ream Thru or Ø .500 Drill Thru.

14. *Call out tap drill depth when important.*

    Tap drill depth may not be an important design issue. When it is, the designer should state the tap drill depth on the drawing as per the following example:

    Ø .201 ▽ .87, 1/4–20 UNC-2B ▽ .63

If it is not important, a simpler call out is sufficient, such as:

$$\text{1/4--20 UNC-2B } \overline{\vee} \text{ .62}$$

The "UNC-2B" designation may not be necessary unless the designer wants to make sure he's getting a class 2 thread.

In many cases a call out such as the following is sufficient:

$$\text{1/4--20 } \overline{\vee} \text{ .62}$$

15. *Keep press fit call outs simple.*

I've bored and reamed countless holes for press fit applications and have come to the conclusion that a light press is all that is needed for most applications. The function of most press fits is to locate and retain two items together. A light press is usually all that is needed to accomplish those two goals.

I don't like to over complicate this issue so I'm going to suggest that a .0005″ interference for a light press and .001″ interference for a "standard" press work well for the majority of jobs that come through a typical shop.

Heavy press fits have the following disadvantages: There is a tendency for parts to gall during assembly and disassembly. Secondly, a heavy press can significantly change the diameters of parts. In the case of a bearing, a heavy press can cause bearing damage. In the case of a bushing, a heavy press can close down the ID of a bushing. Third, a heavy press may not locate items as accurately as a light press because there is a tendency for material to distort and possibly flow unevenly under the high stress needed to make a heavy press fit. Fourth, the stress caused by a heavy press may promote cracking of the parent material.

Fits that turn out to be a little light are usually easily remedied. A little Loctite® around a bearing or bushing is often adequate to hold the item in place.

On drawings, I like to see call outs kept simple and descriptive.

For example, a draftsman could call out a hole as follows:

1.1255±.0002 for slip fit with -501 shaft

1.1245±.0002 for light press fit with -502 bearing

1.1240±.0002 for press fit with -503 bushing

Providing bearings and other mating hardware to a machinist gives him something tangible to work with so he can proceed confidently.

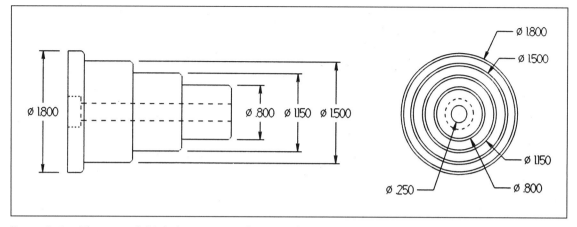

FIGURE 9-1   Diameters of this lathe part are easier to see from a side view drawing.

16. *Avoid getting too carried away with dimensioning circular features with leader lines. (See Fig. 9-1)*

    Multiple leader lines in one view can be difficult to sort out. It is often easier to see circular dimensions in a side view. Note how much easier it is to pick out diameters in the left drawing.

17. *Try to avoid dimensioning to hidden lines.*

    Dimensioning to hidden lines can sometimes make sense in terms of keeping a drawing simple. However, it's not something you should do arbitrarily.

18. *Don't be shy about using reference dimensions on drawings.*

    Double dimensioning is not technically correct; however there are worse errors a draftsman could make, like leaving out a dimension. I've never been hurt by redundant dimensioning, as long as the redundant dimensions were correct. To make double dimensioning technically correct, a draftsman should label the dimension with "(ref.)" following the call out.

19. *Dimension to the centers of the radii of each end of slots, not to the ends of slots.*

20. *Call out "Centers OK" on lathe part drawings.*

    Most long slender lathe parts need at least one end of the part center drilled for tailstock support. Call out "Centers OK" on drawings to remove any doubt as to whether center drilling the ends of a part is acceptable.

21. *Use Delrin when possible for making non-conductive parts.*

Delrin is rigid, moderately priced, holds size well and is easy to machine. Use it whenever possible in your designs instead of other types of plastic. Nylon and UHMW are gummier and more difficult to machine.

22. *Avoid designs that call for press fitting dowel pins into blind holes.*

Protruding dowels sometimes break. Without some provision for removing a broken dowel pin a fix can be difficult. One provision is to have a hole drilled through the part extending from the bottom of the dowel pin hole to the opposite side of the part. The hole allows access to punch out a broken pin.

23. *Give the machine shop some lead time.*

A guy that comes barging into the shop with a hot job demanding action right away is usually not looked upon with much favor from shop personnel. A phone call to the shop beforehand is appreciated and helps people mentally prepare for the onslaught.

24. *Call out fillet dimensions on lathe parts.*

Large shoulder fillets on lathe parts may create assembly problems down the line. If you don't call out shoulder fillet radii on your lathe part drawings and it's left up to the machinist, God only knows what you'll get.

25. *Dimension the overall length of a part in a conspicuous place.*

Avoid putting a dimension on a drawing that looks like an overall length but really isn't. This is a very aggravating thing some designers and draftsman do. They'll put a dimension right across the top of a drawing that jumps out as the overall length. Then later on, after all the raw stock has been cut short, someone will find out it wasn't.

26. *Avoid leaving hidden lines out for clarity.*

This is another aggravating thing some designers and draftsman do. They leave hidden lines out of a drawing thinking they're doing the machinist a favor by not cluttering up the drawing. In actuality it makes the drawing more confusing because a machinist uses hidden lines to check what he thinks he sees in other views. If a hidden line he is expecting to see is not there, the machinist can't be sure he is interpreting the drawing correctly. If you are going to leave hidden lines out of a drawing or view then state it on the drawing.

27. *Return borrowed tools.*

28. *On drawings, provide English equivalents in parentheses following metric dimensions.*

29. *Keep cutter radii in mind when designing parts.*

Small diameter end mills have to be fed slowly since they're not very strong. Design using the largest inside corner radii possible in pockets and cutouts within reason so a machinist can make good time on your job.

30. *Avoid designing parts with knife edges.*

Knife edges are weak, difficult to measure, and confusing to deburr. Unless you are actually designing a knife blade of some kind it is better to truncate knife edges in your designs with a small flat.

31. *Take time to learn about some of the wear characteristics of different metals and their hardnesses.*

Based on what I've seen, proper material selection could save untold hours of machinery down time. Designers need to spend more time learning about and selecting proper materials, hardnesses and coatings to maximize part longevity. One rule of thumb to reduce galling is to use dissimilar metals with dissimilar hardnesses for parts that slide against each other.

32. *Make provision for extracting bearings.*

Many machined parts are used in assemblies for holding and locating bearings. When bearings fail they need to be replaced. I've seen numerous machined parts ruined because maintenance mechanics had no easy way of removing the outer bearing race which is often the only thing left after a bearing falls apart.

A small bearing seat providing access to the outer race from the opposite side allows for easy removal of the race with a punch.

33. *Avoid using gauge numbers to call out sizes.*

Without a chart, gauge numbers mean nothing to the average machinist. Sometimes it can be difficult to locate a chart or even know which one to use since there are so many different standards in the English system. Decimals call outs are the best way to dimension raw materials to avoid confusion.

34. *Flag surfaces that you intend to have machined.*

Knowing when to machine a surface is not always obvious. If the raw stock finish from the supplier is adequate, the designer should state that on the drawing. If the designer

Figure 9–2   Flex type bar clamps work well when designed correctly.

intends for a surface to be machined and it is not obvious, he or she should flag that surface.

35. *Design flex type bar "clamps" that collapse easily. (See Fig. 9-2)*

    These types of bar clamps work great when designed correctly. Some designers don't allow for enough weakness in the flexing area. Without enough weakness, mechanics are forced to apply a lot of torque to the clamping screw which eventually strips the threads.

36. *A chamfered edge is easier to produce than a radiused edge and should be used in your designs whenever possible.*

37. *Make an extra effort to design parts and assemblies that work with relatively loose tolerances.*

    It can't always be done but often it can. In the shop I'm working at now, we have plastic molded parts that have tolerances that are to be held within .001″. It's a constant battle holding those tolerances which insures a lot of down time and creates a lot of stressed out manufacturing personnel such as inspectors, operators, molders, mold makers, and managers. The fault is in the design of the part. One of the hallmarks of well designed parts and assemblies, in my opinion, is that they'll work with relatively loose tolerances.

Rotary Table Magic

I suspect a small percentage of machinists are completely sure of themselves when it comes to setting up and using a rotary table. You'll likely see a wide variety of bizarre methods used by machinists to set up jobs on a rotary table, many of which don't work. Like most things, it is not difficult once you understand it.

To precisely locate a circular part or feature on a rotary table, there are two things you must do.

The first and most important thing you must do is position the part over the true axis of the rotary table. You do that by spinning the rotary table and carefully bumping in the workpiece as you watch the indicator readings. (See Fig. 10-1) When the needle stops deflecting or reads "zero" the axis of the workpiece is directly over the axis of the rotary table. Your indicator does not have to be mounted in the machine spindle to accomplish this, but it can be.

FIGURE 10–1 A part is lined up in the rotary table by spinning the table and bumping in the part until the indicator reads "zero".

The second thing you must do is position the spindle of the milling machine directly over the axis of the workpiece. (See Fig. 10-2) You do that by spinning the indicator around the workpiece and adjusting the X and Y tables of your milling machine until the indicator reading doesn't change or reads zero. The indicator must be mounted in the spindle to accomplish this.

There you go…it's really that simple. I wish someone had explained it to me that way when I was first starting out.

What you'll likely see many people do is position/indicate the machine spindle directly over the hole in the center of the rotary table then position/indicate the workpiece in line with the hole. That method works but is almost never as precise as the first method. One reason it may not be as precise is because the hole in the center of the rotary table is often not exactly concentric with the true axis of the rotary table. Another reason it lacks precision is because by indicating two features to locate the part, (the rotary table hole and the workpiece) you accumulate error.

FIGURE 10-2 The spindle is lined up over the part by moving the machines "X" and "Y" tables.

You can check the concentricity of the hole in a rotary table (if it has one) relative to the true axis of the rotary table by spinning the rotary table while indicating the hole. (See Fig. 10-3) It will likely be off anywhere from a few tenths to as much as a few thousandths depending on the quality of the table.

Bear in mind that when indicating the circumference of a circular object; whether it be in a lathe, mill or rotary table, you get a total indicator reading. (TIR) The TIR of a round part is actually twice the amount of the actual off center error. For example, if you indicate the center hole of your rotary table (assuming it has one) by rotating the table and the total indicator reading is .002" then the hole is really off center of the true axis of the rotary table by half that amount or .001".

You may think "Why should I worry about a few tenths or even a few thousandths for that matter?" You may not have to if you are building bridges but if you are building precision injection molds and other close tolerance tools then it can be important.

FIGURE 10–3 The center hole in this rotary table is checked for concentricity relative to the true axis of the rotary table.

For example, one job that comes up in a mold shop is the task of blending welded areas on existing cores and cavities. This is something that has to be done precisely in order to avoid mismatches. If the core or cavity is round then you can use the rotary table to do the blending. By using the first method explained above, you would be able to precisely locate the workpiece and blend the weld in smoothly.

Locating the workpiece over the center of the rotary table is the difficult part.

After that, the machining techniques used in the rotary table are basically the same as any other type of machining.

Other suggestions that may be helpful are as follows.

1. *Disengage the feed handle to rotate the table quickly.*

    Most rotary tables have a feature that allows the user to disengage the drive gears so that the table can be rotated freely by hand. Turning the table by hand allows you to indicate and position parts quickly.

FIGURE 10–4 Large counter bores are easy to machine with a rotary table.

2. *Disengage the feed handle to machine soft materials. (See Fig. 10-4)*

   The gearing on a rotary table is set up so that by cranking the handle at a moderate pace the table will turn at a rate that is moderate but adequate for most jobs that fit on the table. Bear in mind that for any given handle rotation speed, the larger the part diameter the faster the circumference will be moving. Since you can machine aluminum and other soft materials at high feed rates, sometimes it is easier to rotate the table by hand without the aid of the handle crank or in other words with the crank disengaged.

3. *Machine close tolerance diameters with care.*

   It is more difficult to hold close tolerances on diameters cut with a rotary table than with a lathe.

   Any error you make moving the machine table is doubled on the part. In other words, for each .001″ increment you move your milling machine table, .002″ would theoretically come off the diameter of a part turned in the rotary table.

FIGURE 10-5 A spring loaded scribe can be used to scribe lines on parts mounted in a vise or rotary table.

Another reason it can be difficult to hold close tolerances on the rotary table is because there is a lack of rigidity inherent in stacking that much tooling.

If you have a close tolerance to hold, it is best to approach the dimension carefully. Engage the feed handle and use it. Use a sharp cutter on your final passes and take light cuts at a slow feed rate. Be sure to give the cut ample time to settle in against spring pressure.

4. *Put a reference hole in the center of a square part for easier locating.*

It is more difficult to line up a square feature on center on a rotary table than a round feature. When possible, you should bore a hole or machine a round boss on or in the center of a square part to facilitate the lineup.

5. *Use a spring-loaded scribe to make layout lines. (See Fig. 10-5)*

As I have said before, I am a believer in scribing layout lines to minimize bad cuts, especially for first articles and one-of-a-kind parts.

It is relatively difficult to scribe lines on the face of a round part with a height gauge because there are no flat surfaces to reference to. It may be possible to stabilize the part in a V-block but the setup may be awkward to use.

One easy way to scribe lines on round parts is by using a spring loaded scribe mounted in the machine spindle.

After you locate and clamp your part on the rotary table, you can use the spring loaded scribe to layout either circular or straight lines on the part by moving the part under the scribe. You do that by moving the machines X and Y tables or by cranking the rotary table. The spring loading in the scribe allows light, yet constant layout pressure on the workpiece.

Once you have the layout, it is easy to see what size end mills you can use and what features you need to avoid.

6. *Use the rotary table to machine large counter bores. (See Fig. 10-4)*

How do you machine a large or odd size counter bore in a part? Actually there are a variety of ways. One possibility is to mount the part in a lathe and do it there. Another possibility is to use a boring head that has a facing capability. You could also put the job in a CNC mill and do it there. There probably are other ways but for my money the easiest way is to clamp the part on the rotary table and do it there.

chapter **11** **Taming Warp**

Don't be surprised when parts warp during machining. On the contrary you should expect it. Warp is more of an issue when machining thin and hogged out parts. Thick heavy parts may not require the "special treatment" described in this chapter. Dealing with warp is actually quite simple and becomes second nature after a while. Regardless of how unstable a material is, by following a few of these simple suggestions you will be able to machine parts flat without much difficulty.

Warp is less an issue on lathe parts than on milled parts. Since the material being removed on a lathe part is the same or symmetrical around the axis of the part, the stress or release of stress from the machining is balanced out and the part has less tendency to warp.

Milled parts are more prone to warp. The amount of warpage depends on the internal stresses in the material and the degree to which the parts have thin sections and hogged out areas.

Raw stock that has been cold rolled or cold formed has more tendencies to warp during machining than hot rolled and stress relieved materials.

Whatever the material, you should assume the material will warp to some degree and plan accordingly. The following suggestions should help you make straight, flat parts regardless of how unstable the material is.

1. *Skin cut all surfaces of a piece of material to begin.*

    The first thing you should do is cut through the skin of the material on all surfaces, preferably leaving no surface un-machined.

2. *Let the material warp.*

    Let material "move" between vise clampings. Loosen and tighten the vise between roughing cuts so that the material has a chance to seek its own shape. In other words you want the material to "float" in the vise before clamping it for another cut. What you don't want to do is restrain thin and hogged out parts from the beginning of the machining process to final size. If you do that, when you release the part it will almost certainly bow.

3. *Use shims to prop up plates that need to be clamped directly to a table. (See Fig. 11-1)*

    This suggestion holds true for plates machined on a surface grinder as well as plates clamped directly to a milling machine table. If you start with a warped plate and just pull it down with clamps or a magnetic chuck, the plate will return to its original bow when released. Before clamping down the plate you should measure the gap between the table and the plate in an unrestrained condition. To do that, the ends of the plate must be resting on the table with the bow up in the middle. Before clamping the plate, insert shims or pins on either side of the plate to fill the gap. Then clamp the plate down on the shims and take a cut across the plate. That side will come out flat in an unrestrained condition.

    Toe-clamps work well for clamping down plates this way so that the entire surface of the plate is exposed for machining in one setup. Remember to cut through the skin of the material on both sides before "fine tuning" the plate with the shims.

4. *Tap a plate down on parallels for the last cut only.*

    By letting material float in the vise between cuts or vise clampings, the material will have a chance to seek its own shape and relieve its internal stresses. When you're close to final size on a plate, take a light finishing cut on one side without tapping it down on the parallels to produce one flat surface. Flip the plate over and for the last cut only, tap the material down on the parallels so that the plate comes out parallel. With thin plates, tighten the vise lightly on final cuts to avoid bending the plate.

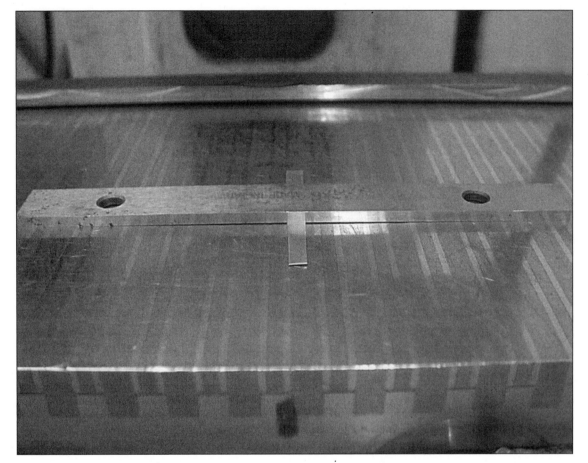

FIGURE 11–1 A warped part is shimmed so that it can't suck down on the magnet. The warp can then be ground out of the part.

A plate machined like this will come out precisely flat and parallel in an unrestrained condition.

5. *Rough out pockets before final sizing.*

Machining deep and/or numerous pockets has a tendency to warp material. To compensate for this tendency you should rough out all pockets first, then go back and finish them and the thickness of the part after the material is done "moving." Leave about .030″ for final cleanup on all surfaces depending on the geometry of the part.

6. *Indicate previously machined surfaces to machine long bars. (See Fig. 11-2)*

To machine long bars in a milling machine and keep them straight there are a few things you can do. First you have to cut through the skin of the bar on opposing surfaces. Do this "skin roughing" quickly disregarding surface finish. Once you've made a

FIGURE 11–2   To cut the warp out of a long bar using a vise, each cut section must be indicated flat before proceeding
with the next cut.

quick cut through the skin over the length of the bar then you can begin creating a
finished flat surface. Do that by taking a long cut with your facing cutter then moving
the bar in the vise. After you move the bar you must indicate the surface you just
machined so that it reads zero along the length of that surface. Then you can cut the
next section while blending in the "Z" heights. What you can't do if you expect to get
a bar flat over its entire length is just slide the bar on parallels as you progress.

7. *Let parts cool before finishing.*

8. *Use a surface plate to check for warp.*

An easy way to check plates for flatness is by holding one end of the part down on a
surface plate and tapping the other end with your fingers. Flip the plate over and
repeat the test. A flat plate won't move when you tap it.

FIGURE 11–3 When indicating long parts on a surface plate, more consistent readings can be obtained by moving the part under the indicator.

9. *To inspect a part on a surface plate, slide the part not the indicator.*

This suggestion is more applicable for checking larger parts. Sliding an indicator mounted on a surface gauge over large areas can sometimes cause the indicator to vibrate and move a little bit. It is better in terms of getting stable, accurate readings to move the part under the indicator. (See Fig. 11-3)

10. *Check your milling machine to see if it is capable of cutting flat surfaces.*

Warped parts can also be caused by a machine that is so worn or loose that it is not capable of cutting flat surfaces.

You can check a milling machine table by clamping an indicator in the spindle and indicating along the length of the table. Start by zeroing the indicator with the table centered over the knee. Then crank the table to each end and watch the readings. If the readings at each end of the table are significantly different then the table is either worn, loose or both. Sometimes you can correct a loose machine simply by tightening the gibs.

chapter **12** **Be Square**

Y ou've probably heard the cliché about building your house on a solid foundation. In machining, square blocks are the foundation of many machined parts. As a machinist, you need to know how to construct precisely square blocks and how to check them. It's nearly impossible to hold tight tolerances on subsequent machining operations if your blocks are not square.

Like many things in machining, there are a variety of methods one can use to construct square blocks.

You may think, "What's the big deal?…every machinist knows how to make square blocks." There's likely more to it than you think.

Blocks

1. *Use a modified ball bearing to hold rough sawed stock in your milling machine vise. (See Fig. 12-2)*

   A steel ball with a flat can be used to apply pressure to the center of the vise and the center of rough sawed stock. The ball should be used when there are no parallel surfaces on the stock for the vise to close on. You may

FIGURE 12–1 Square blocks are the foundation of many machined parts.

have to use the ball a few times to get some surfaces roughly square and parallel before you begin the precision squaring process as described in the following paragraphs. The flat on the ball allows the vise to be closed tightly without damaging or indenting the vise jaw. The ball should be large enough so that it can be held in place without smashing your fingers when you close the vise. A three-quarter or one inch ball bearing ground with a flat about a half-inch in diameter works well.

2. *Face cut one of the widest surfaces first to begin the precision squaring process.*

3. *Cut two sides parallel to continue the squaring process.*

   The main point to remember when squaring blocks is that you have to begin by getting two sides of the block parallel. The best way to get two sides parallel once you've made the first face cut is to set the block high in the vise with the first side resting on some tall parallels. You want to hold the block with as little material as possible so that you can tap the block firmly down on the parallels.

   If you don't set the block high in the vise, you'll have difficulty getting the block to sit firmly on the parallels because as you close the vise, the block will tilt depending on how out of square it is.

   Once you tap the block down on the parallels with a soft hammer, make sure the parallels don't move. If they are snug then you can take the second cut which will make the second side precisely parallel to the first side.

4. *Bury the block deep in the vise to cut the third and fourth sides.*

   Bury the block deep in the vise so the back jaw has plenty of surface area to square the block. Once you square up the third side, it is a simple matter to turn the block over and square the fourth side.

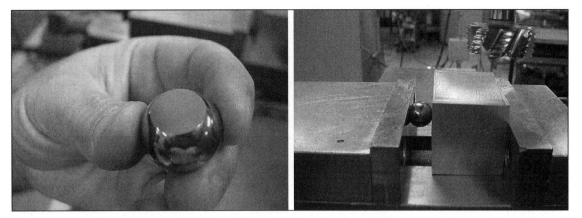

FIGURE 12-2 A modified ball bearing is used to apply pressure to the center of rough sawed stock.

5. *Choose a method for cutting the fifth and sixth sides square.*

Here you have some options.

The most common method is to side mill both ends square while bringing in the overall length. This method is likely the best method for squaring the ends of long slender blocks or bars.

Another option you have is to side mill one end of the block then fly cut the other end.

Still another option that is not commonly used has the advantage of the machinist being able to cut all sides without changing tools. You can use a fly cutter or face cutter to do all six sides. The downside to this method is that you have to make one additional cut and you have to keep track of the orientation of your part or parts.

The technique is as follows: After squaring up the first four sides of the block as per the previous discussion the next thing to do is stand the part up in the vise on one of the rough sawed ends, lock it in place and fly cut the exposed end. Let's call this the fifth side. You need not worry about standing it up straight or square for this cut, just clamp it in the vise approximately straight up and down and take a cut on the exposed end.

Then with your marking pen, place a mark on the upper end of the block on the side facing you. (See Fig. 12-3, left photo) That side and the opposite side are now square to the top of the block. The other two sides are not necessarily square to the top yet because of the part having been set in the vise only approximately straight up and down.

The next operation is to cut the other end square to all sides. You do that by flipping the part over and placing the marked side of the block down in the vise with the mark showing through the side of the vise. (See Fig. 12-3, right photo) Lock the part in the vise. The block should now be standing straight up with all sides square to the bottom

FIGURE 12–3 A block is precisely squared on all six sides using only a fly cutter with the milling machine.

of the vise. Once you take a cut across the exposed end or sixth side, that end will be finished and square to all the vertical sides.

We're still not done yet. We have to go back and recut the fifth side. Remember the fifth side is only square to two sides. In order to cut the fifth side square to all sides you must turn the block over one more time and set the sixth side firmly against the bottom of the vise and take another cut over the fifth side. That is the extra cut I was talking about.

That's it. The block should be precisely square.

The beauty of this method is that you never have to change cutters or use any additional tools such as a machinist's square. Furthermore, your blocks will come out precisely square with little effort and they will have a consistent finish on all six surfaces.

Plates

6. *Square the ends of plates with a fly cutter instead of side milling. (See Fig. 12-4)*

This is a technique that can save you the time and the effort of having to side mill the ends of plates. The technique may seem awkward at first but you'll soon get the hang of it and may grow to like it, especially if you have a substantial number of plates to machine.

First you need to machine two opposite edges of your plates parallel.

Then for the first end cut you need to hold the plate vertically in the middle of the vise up against a machinist's square.

While holding the plate with one hand and the machinist's square with the other you have to figure out a way to tighten the vise. This is where you could use a third hand but since you don't have one you'll have to use your belly, elbow or some other avail-

FIGURE 12-4 A machinist square is used to line up this plate for fly cutting the ends square.

able appendage. Rock the plate back and forth a few times before final tightening to make sure the plate seats itself properly in the vise.

Once you've tightened the vise, you can remove the machinist's square and fly cut the exposed end of the plate. Set the diameter of the flycutter so that you cut the entire width of the plate in one pass.

Use the same procedure to cut the other end of the plate.

Inspection

You may think "What's the big deal?…just lay your square on the part and look for daylight." OK that's one way, but that really doesn't tell you much. A machinist's square will show error but it won't give you a precise or quantitative value of the error.

The most accurate and consistent way to check a block for squareness is by using an indicator and stand that is designed for the purpose. (See Fig. 12-5) I don't believe I have ever seen a stand like this for sale in a tool catalog so you may have to make one. Drawings

FIGURE 12–5 An indicator is used in a special surface gauge to precisely check the squareness of a block. Drawings for the surface gauge are provided in Appendix 1.

are provided in Appendix 1. You may be able to modify an existing surface gauge to accomplish the same thing.

This arrangement works best for checking larger blocks over a couple inches square.

Since this method provides for extreme accuracy, it is often used when grinding blocks. This is a great way for mold makers to check cavity blocks for squareness.

There are two ways you can use this indicator arrangement to check your blocks. With both methods the first thing you must do is adjust your indicator so that there is a few thousandths preload on the point of the indicator with the base of the indicator stand resting against the block.

One method is to adjust your indicator to a zero reading using a precision reference block. You can quickly find "zero" by rotating the indicator base back and forth and watching the arrow on the dial. The maximum throw of the arrow is where you set zero. With the indicator zeroed, you can check your blocks and compare them to the known standard.

The other method is to set the indicator to "zero" with the indicator base up against the block you're inspecting. Then check the opposite side of the block and compare the readings. If the arrow does not return to zero, then your block is out of square by half the amount of that reading at that height up the block.

Bear in mind that although the arrow deflection will increase or decrease depending on how high or low you check the block, the angle of the error remains the same. Therefore, it is best to check the block as high up as you can in order to get maximum arrow deflection, which provides a more accurate reading.

If the indicator reads zero on both sides of the block then you can be sure those sides of the block are precisely square to the base.

chapter 13 Mold Making Tips

Mothers don't let your sons grow up to be mold makers. Plastic injection mold making can make you grow old fast. If you have the temperament, the brains and the skill to build plastic injection molds then you deserve to be paid well for your services. Modern machinery, no doubt, has simplified the mold maker's job; still there is little room for error or sloppy work in mold making.

Although I've worked on many molds and been on teams constructing molds, I don't consider myself an expert. I shy away from mold work simply because I find it too demanding. The way I see it, mold makers live in the worst of all worlds. They work on projects that require the most precision machining on one-of-a-kind expensive parts while being pressured to get things done by people who may not understand the difficulties mold makers face.

Here are a few mold making tips I've collected over the years.

1. *Start with a part print.*

    Seeing a mold drawing package for the first time can be an intimidating experience even for a hardened mold maker. Having a part print to study at the beginning of a project can greatly reduce the mysteries of a new mold drawing package. As a mold maker, you should always ask for and expect a part print.

2. *Continue the drawing review by studying core and cavity details.*

    The core and cavity details are the heart of the mold. Much of the complication you see in an assembly drawing are simply basic components and features such as alignment pins, water lines and ejector system details. Half the battle is understanding how the core and cavity details come together to form the part.

3. *Figure out how the ejector system works.*

    Ejector systems run the gamut in terms of complexity. Getting parts out of the mold can be a challenge for a mold designer, especially when the molded parts have undercuts in them.

4. *Don't worry about going fast when machining molding surfaces, shutoffs (when two or more metal surfaces come together to stop the flow of plastic as it is being injected into the mold) and other close tolerance stuff.*

    The main thing to worry about when machining tight tolerance mold details is getting it right the first time.

5. *Make yourself a small cutter extension. (See Fig. 13-1)*

    This tool is a mold maker's best friend. There are tools commercially available that are similar to this tool but none I have seen are as slim. These extensions work great for reaching into confined spaces and provide excellent visibility which is important when machining small details. This is a tool I use often. Drawings for this tool are provided in Appendix 1.

FIGURE 13-1

This little cutter extension is one of the handiest tools I own. Drawings for the extension are provided in Appendix 1.

6. *Get the lettering correct in your mold cavities. (See Fig. 13-2)*

To get readable words and figures on molded parts, the words and figures on your EDM (Electro Discharge Machine) electrodes have to be readable. In other words, the lettering in mold cavities has to be a mirror image of readable lettering.

7. *Err on the side of making too many electrodes. (See Fig. 13-3)*

Graphite is relatively fast and easy to machine. It is better to have too many electrodes than not enough. Occasionally electrodes break or cavities have to get welded to repair a bad gate or something. Having a few extra electrodes around allows you to re-burn welded cavities without having to use worn electrodes.

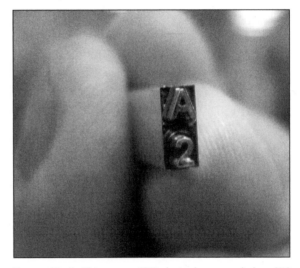

FIGURE 13–2 This copper EDM electrode was made in a CNC milling machine.

8. *Machine graphite with high spindle speeds and moderate chips loads.*

High spindle speeds work best for machining graphite. For the purposes of this discussion I'm going to assume you have a machining center with a "standard" RPM capability. There are special high speed machining centers on the market today dedicated to machining graphite which I know nothing about except that they go fast.

Graphite, as many of you know, is abrasive and should be cut with carbide or diamond coated cutters. HSS wears out quickly in graphite.

Cut graphite dry and keep the cutting area clean by vacuuming out the dust.

Chip loads should be kept in the .003″ to .005″ range for rough milling graphite and the .001″ to .002″ range for finish milling.

FIGURE 13–3 It is better to err on the side of making too many electrodes than not enough.

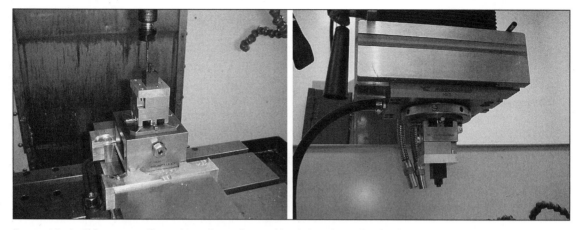

FIGURE 13–4 This system allows electrodes to be machined then immediately placed into service in an electro discharge machine (EDM). This system was made by Hirschmann Engineering.

Chip loads smaller than that will have a tendency to wear tools prematurely. Climb mill to avoid chip out.

An example of a typical cut based on the above conditions would be as follows: You would have to set feed rate at 60″ per minute to get a .0015″ chip load using a four flute end mill spinning at 10,000 RPM.

The formula for determining feed rate, also discussed in chapter 15 is as follows:

Feed Rate (in/min) = Chip Load per tooth x RPM x Number of Cutting Teeth

9. *Scribe accurate layout lines on work before EDM-ing.*

Plunging the electrode in the right location is half the battle when electro-discharge machining. Normally electrode position is determined by using either depth mikes or the machines "beeper" system in combination with digital readout or dial settings. Sometimes you'll be able to position an electrode by sweeping a circular feature with an indicator. If you have that capability, scribing lines is not necessary. Otherwise, I strongly suggest you scribe accurate layout lines on your workpiece to double check electrode position especially when cavities are offset.

10. *Purchase a system for machining and holding electrodes. (See Fig. 13-4)*

The system shown in the photo allows electrodes to be machined in a holding fixture then moved for immediate use in an EDM machine. Once set up, electrodes can be quickly produced and placed into service with precise repeatability using this system.

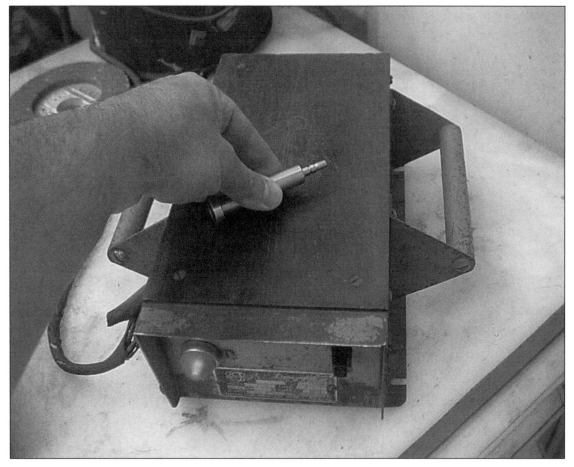

FIGURE 13–5 Thoroughly demagnetizing parts before EDM-ing them reduces the chances of arcing caused by small chips.

11. *Drill flushing holes in deep cavity electrodes.*

   Adequate flushing is necessary for doing high quality electro-discharge machining. Small flushing holes in the .030″ range provide adequate flushing for the majority of small jobs. Drilling deep tiny holes in graphite can be challenging because graphite dust can bind a drill. One way to make the job easier is to start with a large diameter drill bit then use progressively smaller ones as you drill deeper.

12. *Thoroughly demagnetize parts that are going to be EDM-ed. (See Fig. 13-5)*

   Tiny chips that cling to parts can cause arcing. Once arcing starts it can be difficult to stop. Thoroughly demagnetizing parts helps reduce or eliminate arcing.

FIGURE 13-6   Gauge blocks are used to precisely measure the width of a pocket.

13. *Use gauge blocks to precisely measure mold base pockets. (See Fig. 13-6)*

14. *Check features on core pins for concentricity in a V-block. (See Fig. 13-7)*

     Rotate a pin in a V-block to accurately check other circular features on the pin for concentricity.

15. *Check the angle of a single flute cutter by indicating across the cutting edge. (See Fig. 13-8)*

     If you need a precise angle ground on a single flute cutting tool, you have to inspect the angle and make adjustments accordingly. One way to inspect the angle of a single flute tool is to put the tool in a V-block mounted on a sine plate and indicate over the cutting edge. You can also inspect the angle in an optical comparator. Don't rely on grinding machine settings to give you precise angles.

FIGURE 13–7 A circular feature is checked for concentricity by rotating the part in a V-block while indicating the feature.

16. *Ream ejector pin holes from the back sides of cavities to avoid bell-mouthing.*

    Bell mouthed ejector pin holes are prone to flash. There is a tendency for a reamer to cut a little over size when it first starts into a hole which is what causes bell mouthing. Reamed holes are more accurate on the exit side.

17. *Drill and ream ejector pin holes after rough polishing molding surfaces.*

18. *Rough polish molding surfaces before heat treating.*

19. *Finish polish molding surfaces after heat treating.*

20. *Machining core and cavity details is faster but riskier than EDM-ing them.*

21. *EDM-ing core and cavity details is safer but slower than machining them.*

FIGURE 13-8 The angle on this single flute cutter is being checked for accuracy using a sine plate and indicator.

22. *On mold details as well as other machined parts, try to err on the side of being metal safe.*

23. *The zero corner of mold plates should be stamped or located next to the offset leader pin.*

24. *Grind cavity vents the safe way.*

    When grinding the deeper secondary relief vents which are usually about .005″ deep, it is safer to plunge the grinding wheel down near the cavity and then feed away from the cavity. If you feed your wheel toward the cavity you may get into trouble by either inadvertently breaking into the cavity or by burning the vent as you slow your table feed to avoid the cavity.

25. *Feed away from your spin fixture when grinding pin diameters. (See Fig. 13-9)*

    The leading edge of a grinding wheel always breaks down first. By plunging the grinding wheel down and feeding it away from the spin fixture, you will always have the

FIGURE 13-9 The OD of a pin is being ground in a spin fixture.

lowest portion of the wheel in contact with the pin which enables you to grind pins parallel up to shoulders.

26. *Number and identify mold components for easy tear down and assembly.*

27. *Don't skimp on machining pry bar slots in mold plates.*

Pry bar slots are something that should be liberally designed into a mold. They make mold tear down easier. An eighth inch deep pry bar slot is sufficient in most cases.

28. *Leave stock on the back side of your core and cavity blocks.*

If you leave stock on the back side of your cavity blocks during the initial phases of machining, you'll have stock to work with in case you need to machine deeper in order to repair a bad cut. If you make cavity blocks exactly to size in the beginning and you make a bad cut, you'll probably have to get the mistake welded. Once you get a good cavity, you can whack the extra material off the back side.

FIGURE 13–10   Final polishing with Simichrome® produces a high polish.

29. *Plan on cutting core and cavity details a few thousandths deep in the beginning so you can polish up to the parting line.*

If you polish up to the parting line on a cavity that has been cut exactly to depth, it's hard to avoid rolling the edge of the parting line. By cutting the cavity a few thousandths deeper in the beginning (usually .005″) you'll have stock to grind to remove any rolled edge. That way you'll end up with a sharp parting line.

30. *Use balsa wood and diamond compound for finish polishing.*

Balsa wood impregnated with diamond compound cuts quickly and leaves an excellent polished surface.

31. *Use Simichrome® paste for final high luster polishing. (See Fig. 13-10)*

Because of its very fine abrasive, Simichrome® works great for polishing when you want to avoid scratching a surface. Use a wooden stick in conjunction with the paste to apply pressure to the workpiece. Final polishing with paper or cloth will produce a high luster.

32. *Find a good welder.*

It takes a skilled welder to weld core and cavity details without warping the material, creating voids and excessive sink. A welder needs to keep as much heat out of parts as possible. That may involve patiently welding in thin layers and letting the part cool between each layer.

FIGURE 13–11   It's good practice to leave a little weld showing when blending weld on mold parts.

33. *Leave a little weld showing when blending weld on a parting line or molding surface. (See Fig. 13-11)*

    Since welds are usually a fix or patch, it's good practice to avoid grinding or machining into the original or parent material. Leaving a weld a tenth or two high (.0001″–.0002″) usually has no noticeable affect on mold function or part appearance.

34. *Use hardened pins when lapping to avoid galling.*

    Occasionally a hole will need a little lapping to bring it to final size. Lapping should be used when there is only a few tenths of material to remove. I prefer lapping a hole with a close fitting dowel pin instead of an adjustable lap to avoid the possibility of bell mouthing. Soft materials gall easily. It is best to use a hardened dowel pin or ejector pin to lap with to reduce the risk of galling. Use a lot of oil when lapping and very slow spindle speeds to avoid generating heat or drying out the hole. Hand lap when possible. A pin galled in a hole can be difficult if not impossible to remove without ruining the work.

FIGURE 13–12   A split bushing can be used to hold ejector and core pins for easy insertion in a lathe.

35. *Make a bushing to slide over core pins and ejector pins for easy handling in a lathe. (See Fig. 13-12)*

This is a simple and effective way to save time. When you're working on a number of core pins, the head of the pin can make it difficult to get the pins in and out of a lathe quickly. One way to get them in and out easily is by holding them with a bushing. Otherwise, you'll have to remove the collet each time you want to change pins or you'll have to crank the chuck in and out an excessive amount to clear the head of the pin.

The OD of the bushing needs to be slightly larger in diameter than the head of the pin. Be sure to make the OD a standard collet size. You can make these types of bushings out of a hard T6 type aluminum but just about anything will work.

You can cut slits in the bushing with a band saw or hack saw since the size of the slit is not critical within reason. One side of the bushing must be cut through and the opposite side of the bushing may need a partial slit in order to flex easily. It is best to make the partial slit from the outside of the bushing so that you don't lose too much material from the ID of the bushing.

36. *Use silicon sealant to seal corroded O-ring seats.*

Molds that have been in production for a while may develop corrosion around or underneath O-ring seats. Dab a little silicon on the corroded areas during assembly to insure against a water leak.

37. *Be careful about inadvertently pulling an ejector plate out of a mold.*

Once you pull ejector pins out of their holes when mounted in the ejector plate it can be difficult or impossible to get them back in without disassembling the plate. Ejector pins go in best one pin at a time.

38. *If you happen to ding a molding surface don't automatically stone off the resulting crater.*

Often times the crater from a dent can be carefully peened back into place. Once peened, you may be able to polish out the remaining blemish.

Get Your Grinder Goin'

Surface grinders are neat machines. A grinder, like no other machine, can transform a hard, crummy looking piece of steel into a shiny, functional part. When you walk into a grinding room, you may get the feeling you're walking into a different world. Grinding is usually the last and most precise operation performed on a part before it is placed into service. It often requires more care and concentration than other types of machining.

The following is a list of suggestions that may make your time in the grinding room a little easier and more productive.

FIGURE 14–1 Dykem® is used as a visual aid to blend in weld.

1. *Mount a surface grinding wheel firmly.*

   Use paper blotters on both sides of a grinding wheel and use a little "umpf" when tightening the jam nut so the wheel won't shift during a spindle startup or heavy cut.

2. *Stop and start a surface grinding wheel as often as you like.*

   One myth you may have heard is that you should never turn a grinding wheel off and on without re-dressing it. As long as you tighten the wheel sufficiently, you can stop and start the spindle as much as you like without re-dressing it.

3. *Stop the wheel to pick up a surface.*

   It is not necessary to stop a wheel to pick up a surface but it is usually faster and safer. Cranking a spinning wheel down while eyeballing the gap between the wheel and part can be tedious. If you happen to crank too far you may gouge the work.

   If you turn the spindle off, you can approach the part faster because there is no risk of gouging the work. Once you touch the part with the wheel you can set a "rough" zero. After backing the wheel away, you can turn it on and re-approach the wheel to the work within a few thousandths with confidence. Then you can carefully touch the part with the wheel running and reset a more precise zero.

4. *Spread a thin layer of Dykem® on a surface to locate the surface. (See Fig. 14-1)*

I've heard it said that a layer of Dykem is about .0002″ thick. From what I've seen the thickness of a layer of Dykem can vary considerably. Nevertheless, by applying a thin layer to a surface you'll be able to see the Dykem being removed by the grinding wheel before it touches the metal surface. This is a great way to blend a weld to an existing surface. The photo shows a weld that was ground flush using this method.

5. *Err on the side of using softer wheels for heavy stock removal.*

I once heard someone say "use a soft wheel for grinding hard material and a hard wheel for grinding soft material." At the time it sounded like good advice. After some trial and error, I came to a different conclusion which goes like this: "Use a soft wheel for grinding hard material and soft wheel for grinding soft material." Either way you'll find that grinding soft or gummy material is considerably more difficult than grinding hard material.

Forty-six grit soft wheels in the "H-I" hardness range work well for flat grinding and heavy stock removal of most materials. Wheel hardnesses are given by a letter code shown on most wheels ranging from "A" through "Z" with "A" being the softest. Grades "H" through "K" are commonly seen in shops.

6. *Use porous wheels for heavy stock removal. (See Fig. 14-2)*

When you walk into a grinding room and look at the wheels that have been dressed down the farthest, you'll notice that the majority of them will be the forty-six grit porous wheels. What does that tell you? It tells you people use them because they work. Wheel porosity or structure is given by a number code shown on the wheel ranging from 1 to 16, with 1 being the most porous. Structures in the 8 through 12 ranges are commonly seen in shops.

FIGURE 14-2 Porous grinding wheels have less tendency to load up and burn surfaces than non-porous wheels.

FIGURE 14–3 Heavy cuts can be made with a surface grinder if material is fed in a little at a time.

7. *Rough off stock in a surface grinder by taking deep, narrow cuts. (See Fig. 14-3)*

   You can take relatively heavy cuts with a grinding wheel if you feed the wheel across the part a little at a time such as .005″ to .015″. If you feed much more than that your wheel will break down prematurely. The photo shows a .030″ deep cut using this method. Use a soft, porous, 46 grit wheel to avoid wheel loading, burning and glazing.

8. *Use harder, finer grit wheels for form grinding.*

   Soft, porous wheels are better for flat grinding and heavy stock removal since they won't burn material or load up too easily. To maintain sharp corners and small radii such as when form grinding, finer, harder wheels work better. Sixty and eighty grit wheels in the "J-K" hardness ranges work well.

FIGURE 14-4 Thin parts can be blocked in for surface grinding using common radius gauges.

9. *Use slow feeds to produce smooth, accurate surfaces.*

   Feed rate, depth of cut, and grinding wheel condition have a lot to do with surface finish. To produce smooth finishes you should dress the wheel, take very light cuts such as .0001″ and feed slowly.

10. *Use a set of radius gauges to block in small, thin parts on a magnetic chuck. (See Fig. 14-4)*

FIGURE 14–5 An indicator is used to compare the height of a part to a stack of gauge blocks.

11. *Be careful when using depth mikes to measure down to a magnetic chuck.*

    If there is a little slop in the threads of your depth mikes, a magnetic chuck will suck the tip of the mike down and may give you a false reading. You can overcome the error by continuing to turn the thimble until there is a little backpressure.

    It is safer in terms of getting a precise measurement to remove the part from the magnet and use either outside mikes or an indicator to compare the part to a stack of gauge blocks. (See Fig. 14-5)

12. *Set a block on dowel pins when using an angle plate to square an edge. (See Fig. 14-6)*

    By setting a block on dowel pins you avoid the risk of sucking the block down on a non-square edge and possibly tilting the angle plate. Once one edge is square the plate can be flipped over and ground on the opposite side without using dowel pins.

FIGURE 14-6 A block is supported on dowel pins to keep a non-square edge from sucking down on the magnet.

13. *When grinding pins, blend angles and straight sections at he same time. (See Fig. 14-7)*

Core pins that have angles and straight sections blended together are commonly used in molds. One example of such a pin would be one needed to form a countersunk screw hole in a molded part. The easiest way to grind a pin like this is to dress both an angle and straight section into the grinding wheel. You can plunge the wheel straight down to form the diameter and the angle at the same time while rotating the pin in a spin fixture. To create a sharp corner at the vertex of the angle and straight section, the wheel should be redressed before final sizing.

14. *Dress angles on grinding wheels using preexisting angles. (See Fig. 14-8)*

Instead of using a fancy dressing tool for dressing angles on a grinding wheel you can use preexisting angles and a diamond nib holder. Set the preexisting angle next to a small angle plate for support then manually slide the diamond across the center of

FIGURE 14–7 An angle and straight section are ground at the same time on this pin.

the grinding wheel using the preexisting angle as a guide. The dresser in the right hand photo can be used to dress precise angles and radii on grinding wheels.

15. *Use a magnetic sine plate for grinding angles. (See Fig. 14-9)*

16. *Use a wet towel to cool parts before measuring.*

Grinding generates heat and heat expands parts. To hold accurate length measurements on long parts such as core pins and ejector pins they need to be about room temperature. You can either let them cool gradually or you can help them cool by wrapping a wet towel around them momentarily.

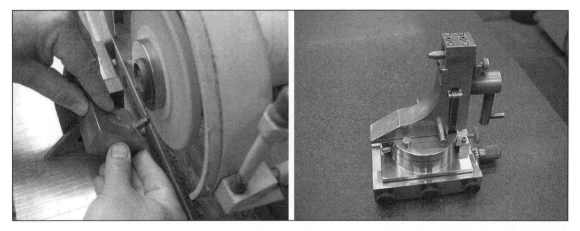

FIGURE 14–8 Angles can be dressed into grinding wheels using preexisting angles to guide the diamond. Precision angles and radii can be dressed into grinding wheels using fancy wheel dressers such as the one shown on the right.

FIGURE 14–9 Magnetic sine plates provide an easy means for holding and grinding angles with a surface grinder.

FIGURE 14-10   The OD of slender pins need to be ground with the aid of a support. Two different types of supports are shown here.

17.   *Use a thin, fine wheel for grinding weak, slender diameters. (See Fig. 14-10)*

Grinding the diameters of long slender pins can be challenging because there is a tendency for slender pins to vibrate and push away. Dress a fine grit wheel with a tiny cutting area to reduce grinding pressure. An .050″ or thinner cutting area dressed into a sixty or eighty grit wheel works well. The use of an end support will be needed to reduce flexing and vibration. The photos show two different types of end supports. The end support in the left photo was made with banding material. A small V-groove ground in the end of the band is used to support the pin. The support in the right photo is a more versatile tool and uses an adjustable spring loaded plunger to provide support.

18.   *Place a thin piece of phenolic on a magnetic chuck to aid sliding a surface gauge on and off the magnet. (See Fig. 14-11)*

Sometimes it can be handy to use a surface gauge directly on an engaged magnetic chuck. A thin piece of phenolic, cardboard or a few layers of paper between the chuck and the surface gauge reduce the magnetic force on the gauge so that it can be slid in and out of position.

19.   *Make a wheel dresser to use for your pedestal grinder that won't cover you with grit. (See Fig.14-12)*

I made one of these tools years ago and use it for dressing pedestal grinding wheels. It provides good leverage and allows you to stand off to the side of the wheel. You can make one of these dressers by installing a set of dressing stars on the end of a bar with a shoulder screw. The stars need to rotate freely.

FIGURE 14-11   A thin piece of cardboard, plastic or phenolic can be placed over the magnet to allow a surface gauge to be slid into place.

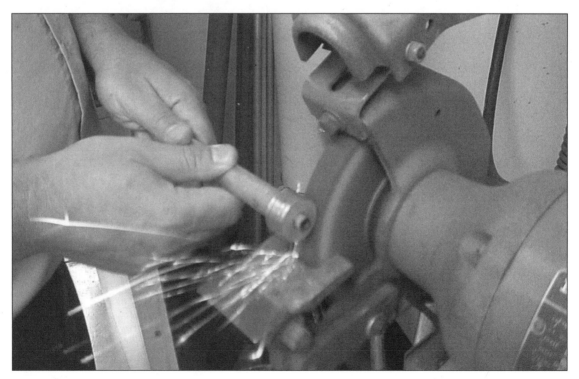

FIGURE 14-12   A simple wheel dresser for a pedestal grinder can be made by mounting a set of dressing stars on the end of a bar.

FIGURE 14–13   A Norbide® stone is used to dress clearance into a grinding wheel.

20. *Use a "Norbide®" stone for offhand wheel dressing. (See Fig. 14-13)*

These stones can be used to quickly dress clearance and shapes on wheels. It is a very useful tool I use often when grinding.

21. *Make a diamond nib holder that can be used to dress the side of a wheel. (See Fig. 14-14)*

Angle the nib in the holder so that it can be rotated to avoid wearing the diamond in the same spot.

FIGURE 14–14   The side of a wheel is being dressed with a diamond nib.

22. *Use an adjustable wheel dresser for reaching wheels in various positions. (See Fig. 14-15)*

These dressers are versatile tools that can save you time. Note the two diamonds mounted on the dresser. One is for dressing the side of a wheel and the other is for dressing the bottom of a wheel.

23. *Use a fence to accurately locate work on a magnetic chuck. (See Fig. 14-16)*

This is a tool that can be used to accurately locate work in the "X" direction. The fence has a slot that slides over the rail and is anchored to the rail with three screws.

FIGURE 14–15   This versatile wheel dresser has plenty of range and is capable of dressing the sides and bottoms of grinding wheels in a surface grinder.

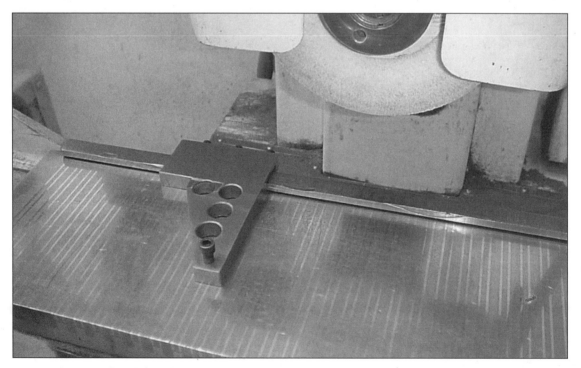

FIGURE 14–16   A "fence" that clamps to the back rail allows work to be precisely located on a magnetic chuck.

FIGURE 14–17   A thin contact area dressed into the face of a grinding wheel used for "side wheeling" reduces cutting pressure and wheel glazing.

24. *Dress a grinding wheel by feeding the diamond across the wheel fairly aggressively.*

Feed the diamond at a fairly brisk pace across the wheel to rip out the abrasive crystals instead of dulling them. Once you have gone across the wheel briskly the first time, then go back and forth across the wheel a few times at a slower pace to make sure there are no dangling crystals.

25. *Climb grind when finishing forms such as radii and slots to produce smooth, accurate surfaces.*

Conventional grinding with a form wheel throws sparks and heat into material that has yet to be ground which may cause slight material distortion. Climb grinding throws sparks and heat over an area that has already been ground. You can get a slightly better finish by climb grinding with a form wheel on final passes.

26. *Dress a thin cutting area into a wheel when "side wheeling." (See Fig. 14-17)*

"Side wheeling" can be difficult because there is a tendency for the wheel to glaze which is a result of the relatively large wheel area that is in contact with the work. To

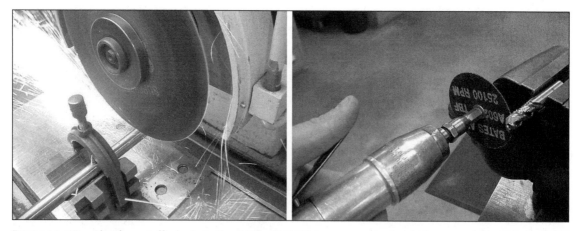

FIGURE 14–18   Abrasive cutoff wheels in the .060"–.070" thick range can be fed fairly aggressively. Thinner cutoff wheels, such as the one shown on the right, work well in hand-held spindles.

reduce glazing, dress a thin contact area into the wheel. .030" to .050" wide usually does the trick. If the wheel still glazes then you may need to chip out some areas of the wheel that contact the work so that the cutting action is interrupted. Use coarse, friable (brittle and crumbles with relative ease) wheels to help reduce glazing.

27. *Disregard wheel color when selecting grinding wheels.*

Wheel color is virtually meaningless.

28. *Avoid using extremely thin cutoff wheels. (See Fig. 14-18)*

Abrasive cutoff wheels under a 1/16" thick flex a lot under load. Wheels in the .070" thick range work well for most parting off jobs. Thinner wheels work well in hand held air spindles for cutting off small pins. Unlike many machinists, I prefer feeding a cutoff wheel down in the "Z" direction with the hand crank rather than moving the stock into the wheel in the "X" direction. I believe you can get a better feel for the amount of load you put on a cutoff wheel if you feed down on the stock.

29. *What does friable mean?*

Friable means brittle. Friable abrasives break easily forming new sharp edges. In a grinding wheel there are two things that break down. One is the abrasives themselves and the other is the binder. For best results, both should break down at about the same rate.

30. *What is the difference between aluminum oxide and silicon carbide?*

Silicon carbide crystals are harder and fracture more easily than aluminum oxide crystals. Silicon carbide wheels are used to grind hard, brittle material such as carbide,

FIGURE 14-19   Blocks can be set at an angle on a magnet to avoid concentrating heat in one area during grinding which may help reduce heat distortion.

stone and ceramic. The common "green" wheels used in pedestal grinders for grinding carbide consist of silicon carbide crystals. Aluminum oxide crystals are tough and resist fracture. They are gradually dulled by hard material. When grinding pressure is great enough they will eventually fracture to present new cutting edges. Aluminum oxide wheels are identified by the letter "A" and silicon carbide wheels by the letter "C" within the wheel designation shown on the side of most wheels.

31. *What is the difference between vitrified bonding and resinoid bonding.*

Vitrified bonding material is glass like and brittle. Vitrified bonded wheels are commonly used in surface grinders. Resinoid bonding material is flexible which enables it to withstand harsh treatment. Resinoid bonded wheels are commonly used in pedestal grinders for off hand grinding.

32. *What's the purpose of grinding on the skew? (See Fig. 14-19)*

You'll occasionally see machinists set blocks at an angle on the magnet when surface grinding. Heat and warp are reduced using this technique as a result of the shorter contact time the wheel has with the part and the fact that heat is not being concentrated in one area. The technique is more effective on long thin parts that have a tendency to suck or expand into the grinding wheel from heat buildup.

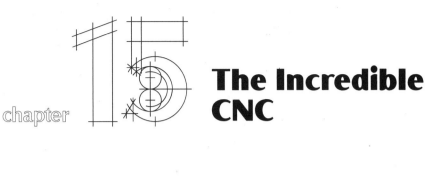

chapter 15

The Incredible CNC

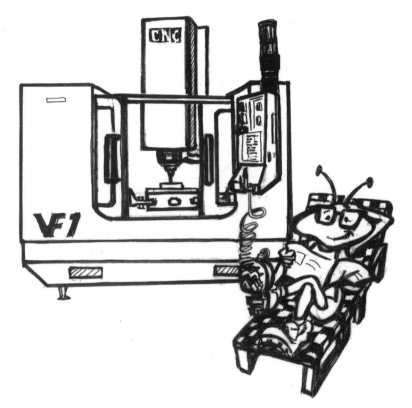

When I first started machining, CNC machines were just coming into mainstream use. The first few shops I worked in had only conventional machines. It wasn't until the mid 1980's I started to notice the incredible versatility of CNC machines. The contouring and shapes the machines could generate made many jobs much easier. The use of templates to file and sweep in surfaces was rapidly becoming a thing of the past.

Initially, I was intimidated by the technology and couldn't imagine how a person could machine parts simply by pushing buttons on a control panel. The mysterious code that ran the machines seemed to be the domain of computer people and was seemingly beyond my comprehension. It wasn't until later when I took a class in programming that the mysteries started quickly dissolving. Taking a class in programming is a great way to get started in CNC machining.

As versatile as CNC machines are, they are only as good as the people programming and operating them. The cliché "garbage in garbage out" is well suited to CNC programming and machining. There is simply no substitute for proper planning and machining know-how. A programmer without much machining experience would likely struggle to produce good parts consistently.

One of the great virtues of CNC machines is that while they are running, the machinist or operator can be doing other things. In essence, they take the labor out of machining.

Another virtue of CNC machines is that the cutting and measuring process so prevalent in conventional machining is virtually eliminated. If programmed and setup correctly the cutter will go precisely where it should so that dimensional accuracy comes quickly. Usually only minor adjustments are needed to compensate for slight variations in tool size.

For anybody that wants to learn the machining trade, I believe it is best to concentrate on conventional machining in the beginning. There is a certain machining intuition that is best developed on conventional machines. Having said that, I see nothing wrong with learning the computer side of machining as you gain experience on conventional machines.

The following are some suggestions that may help you with your CNC machining projects.

1. *Learn a CAM (computer aided machining) program. (See Fig. 15-1)*

   CAM software, once you are familiar with it, can make CNC programming incredibly easy. It is important to also learn manual programming in order to efficiently edit programs at the machine.

2. *Before writing a program, check your end mill inventory.*

   It can be frustrating to write a program then find out you don't have the end mills you selected for the program.

3. *Set speeds and feeds conservatively in the beginning. You can always bump them up once you get some history on how things run.*

4. *Baby-sit virgin programs.*

   It is nearly impossible to write a program that needs no editing unless it is a simple program. Not until you watch the machine go through the motions and cut a part will you know for sure how things are going to run. It's good practice to keep your finger on the feed hold button when proofing programs to avoid crashes. Another thing you can do to be safe when proofing a program is to run the program with the tools high above the part.

FIGURE 15-1 CAD systems are increasingly making their way to the shop floor. This system is available for anybody in our shop to use.

5. *Count on destroying a least one setup part.*

    On complex parts it is difficult but not impossible to get everything right the first time through the editing and debugging process. It's a good idea to have a setup part made of some soft material such as aluminum or wax to work with in the beginning so that you can forge ahead confidently without worrying about scrapping a part.

6. *Program in short segments.*

    Short program segments are easier to edit and work with than long endless streams of data. I prefer separate program segments for each feature of a part within reason so that in case a feature needs to be re-run, the programming for that feature can be quickly found, edited and re-run. Use canned cycles when possible.

7. *Use the mirror function on your controller instead of generating additional programming.*

In many instances, especially when you have right and left hand parts, the mirror function on your controller can be used to quickly provide tool paths for opposite parts and features. The function can save programming and editing time. Be aware that when you mirror something, climb cuts become conventional cuts and vice versa. You may have to go back to the computer to switch the cutting method if it is important.

8. *Program to reduce burrs.*

CNC machines can produce parts so quickly it sometimes takes longer to deburr a part than it takes to machine it. Take a little extra time to program so that burrs are reduced. Program end mills to climb mill into edges on final cuts. Leave about .005-.007″ material on your final cuts so that large burrs get cut away. Also program with separate end mills for roughing and finishing so that your finishing end mills stay sharp which helps reduce burrs.

9. *In the mill, be especially careful when clamps, bolts and other hardware protrude above "Z" zero.*

Insufficient retract heights during lateral movements are a common cause of "crashes" in CNC machines. When you have protruding hardware, you have to be especially careful with your programming to avoid these crashes. Use G98 in canned drilling cycles. G98 returns the spindle to the initial plane which should be set high enough to clear any protruding hardware for lateral moves. G99 returns the tool to the "R" plane which is commonly set close to the top of the workpiece.

10. *Keep peck drilling distances short so that drill bits don't get clogged with chips.*

A tenth of an inch peck usually works well on drills in the ¼″ range and larger. You can increase the pecking length if chips are naturally breaking into small pieces. On smaller diameter drills a shorter peck is advisable to clear chips more often.

11. *Reduce cycle times by:*

- Minimizing tool changes.
- Staging tools for undelayed tool changes.
- Avoiding "air" cutting by programming with minimal tool runoff.
- Reducing depth of cut while increasing feed.

12. *Input different feed rates for "Z" moves and lateral moves.*

Programming different feed rates for "Z" moves and lateral moves gives you better control over the program. You'll be able to do "global" editing at the machine for

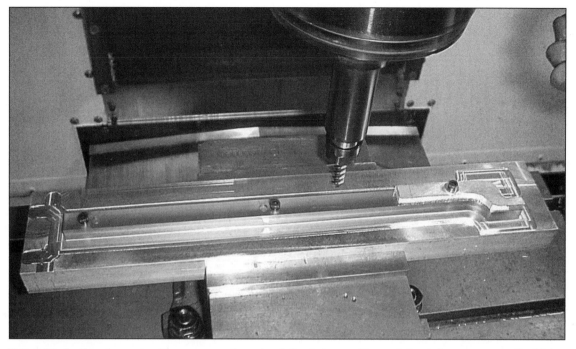

FIGURE 15–2 Clamping parts with cap screws through existing holes provides unobstructed access to parts for further machining.

just the moves you want to change. If all your feed rates are initially the same then each time you do a global edit everything will change which may not be what you want. It is also good practice to input different feed rates for different tools for the same reason.

13. *Use separate tools for roughing and finishing.*

This is an important suggestion for consistently producing quality parts. In the long run, parts and features cut with separate tools for roughing and finishing will look better and be more accurate.

14. *Design fixtures with plenty of room for tool runoff.*

When designing fixtures to run multiple parts, make sure you leave enough room between parts so that cutters don't interfere with adjacent parts as they lead on and off.

15. *Use thru holes in parts for clamping. (See Fig. 15-2)*

Strap clamps can be awkward to use when they don't allow complete access to a part for machining.

Many parts are designed with thru holes or slots in them that can be used to both locate and hold parts for further machining. One of the problems you'll face using

FIGURE 15–3 Care must be taken when cutting out a "window" so the resulting slug doesn't jam the end mill. In this instance the majority of the slug was drilled out to avoid that problem.

thru holes to locate parts is that many holes in parts are 1/64″ to 1/32″ over some nominal screw size. Since the hole is oversize you can't very well use the corresponding screw to accurately locate the part.

One way you can overcome that problem is as follows: Turn the threads of the next larger screw down to match the size of the clearance hole or slot. For example, a clearance hole for a #8 screw would probably be about .180″ in diameter. If you take a #10 screw which is about .187″ in diameter, you can turn the threads down on the screw to just under .180″ and use it to locate the part using the #8 clearance hole. What's left of the turned down threads on the #10 screw will still have more than enough strength to hold the part securely.

Another way to locate parts using clearance holes is to use flat head screws. The taper on the flat head screw will center a hole around the screw. This method may not be as accurate or as rigid as the above method since flat head screws are not always precisely made and they don't provide much surface area for retaining parts.

16. *Program to drill out islands and slugs when practical. (See Fig. 15-3)*

If you are milling a window of some sort completely through a plate, it is best to drill out the center section of the window when practical so that very little or no slug remains. If the center section of a window is not drilled out, the slug may jam against the end mill and possibly break it as the slugs starts to come loose from the part.

If the window is so large as to be impractical to drill out, then you may be able to clamp the window in place so that it won't move when it is cut through.

FIGURE 15–4 A refractometer can be used to check mixture ratios of water and soluble oil.

17. *Clear tool diameter offset registers after finishing a job.*

   With so many numbers floating around inside a CNC controller, it is easy to forget to clear offset and tool wear registers after running a job. That can be costly in terms of ruining parts on upcoming jobs.

18. *Clear unused tool length offset registers.*

   Occasionally during the course of programming or editing an improper "H" value which calls up a tool length offset will find its way into the program. If the improper "H" value calls up a tool length offset number that has a larger "Z" negative value than the tool in the spindle is supposed to have, a crash may occur. Set all unused tool length offset registers to zero to minimize the risk of plunging a cutter too far.

19. *Maintain correct coolant ratios.*

   When we first got our CNC machines we used the "that looks about right" method for mixing the soluble oil and coolant water. Things started rusting so we poured in more soluble oil which stopped the rusting but created a slime problem. It wasn't until we bought a refractometer to measure the ratio that we got it right. (See Fig. 15-4) As it turns out the "no rust, no slime" window is fairly narrow and can be difficult to find without the use of a refractometer.

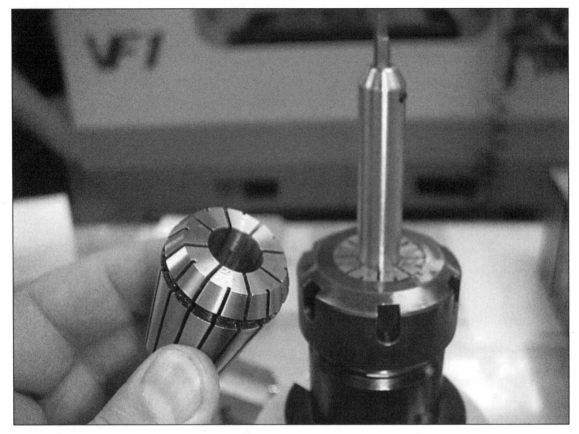

FIGURE 15–5 ER collets are versatile because they have a .030″ tolerance range which allows them to be used to hold various diameter cutters and drill bits.

20. *Avoid withdrawing reamers with a rapid move.*

By virtue of the close fit a reamer has with a hole, it is safer to extract a reamer using a slow feed rate or at the same rate used to ream the hole in the first place. Jerking a reamer out of a hole with a rapid move can cause the reamer to bind in the hole which may either break the reamer or pull the workpiece out of the vise.

21. *Avoid using piloted tools in CNC machines.*

Piloted tools such as counter bores work great in drill presses because of the quick alignment they provide. There is no advantage to using piloted tools in a CNC machine since lineup is not an issue. End mills with bottoms ground square and relieved to form cutting edges work fine as counter bores. Even standard end mills with their little bit of end relief can be used as counter bores in most cases.

The disadvantage of using piloted tools in a CNC machine is that they create confusion as to where to set "Z" zero. If "Z" zero is set at the cutting edges which would make the most sense then special care must be taken to provide adequate "R" plane clearance so that when the table moves the pilot isn't still in a hole.

FIGURE 15–6 A dowel pin is used in a tool station and programmed to create a work stop in a CNC mill.

22. *Use stick wax to lubricate taps in CNC machines.*

Stick wax can be used to reduce tap failure and resists being washed away in a coolant stream.

23. *Choose a collet system with a large tolerance window. (See Fig. 15-5)*

"ER" collets have about a 1/32″ tolerance window which makes them ideal for holding odd size diameters rigidly and accurately. Collet systems such as the R8 system used in Bridgeport mills and the 5C system have narrow tolerance windows that don't allow for holding off-size diameters rigidly and accurately. A 5C or R8 collet squeezed down to hold an undersize diameter will not hold the diameter with much accuracy or stability.

24. *Use a dowel pin in a tool station to create a stop. (See Fig. 15-6)*

I've seen machinists program a tool holder holding a dowel pin to create a workstop. I've never adopted the practice but I see nothing wrong with it.

FIGURE 15–7 A 1-2-3 block is used as a removable stop so that both ends of the block can be machined in one setup.

25. *Use a stop that can be removed so that both ends of a part are accessible for machining in one setup. (See Fig. 15-7)*

   The photo shows a 1-2-3 block being used as a removable stop so that both ends of the block can be machined.

26. *Use a lathe to make lathe parts and a mill to make mill parts.*

   Although the suggestion sounds simplistic, I've seen many parts made in the wrong machine which is not very efficient.

27. *Use "numbers" to verify cutting conditons.*

   Two important numbers to look at are "Chip Load" and "SFM." (Surface feet per minute).

Sensory feedback for gauging machining parameters is a little harder to come by with CNC machines than with conventional machines. You're limited to sound, vibration, finish and possibly a spindle load meter for feedback on how things are cutting. Chip color is of little importance in a CNC since the workpiece is usually being flooded with coolant. Also, you can't feel the pressure of the cut like you can with a conventional machine. Therefore the "numbers" become a somewhat more important indicator of correct feeds and speeds.

I usually set feeds and speeds intuitively first then I may experiment to see what I can get away with. Later on I may run some "numbers" to see what's really going on.

28. *Use "Chip Load" to determine the aggressiveness of a cut.*

The formula for determining chip load is as follows:

$$\text{Chip Load} = \text{Feed Rate (in/min)} / (\text{RPM} \times \text{Number of Cutting Teeth})$$

The above chip load formula, though applicable for lathe jobs is more useful for determining chip loads on milling machine cutters.

Chip loading can vary considerably depending on the rigidity of the setup, the strength of the cutting tool and the finish you want. With corncob type roughing cutters you can usually get away with higher chip loads than you can with standard end mills. There is really no right or wrong chip load, none-the-less extremely small chip loads are usually no good because they promote friction, chatter and slow cycle times. On the other hand, you may have to use light chip loads when machining flimsy parts or when using weak cutters.

In a lathe, I consider a chip load in the .020″ per revolution range when cutting steel to be a relatively aggressive roughing cut. With a rigid setup, if you keep your depth of cut shallow, you can get away with it.

In a milling machine, I consider a chip load in the .010″ range when cutting steel to be a relatively aggressive roughing cut. Normally I wouldn't use a chip load that heavy unless I'm roughing using a strong cutter with a rigid setup and taking relatively shallow cuts. Much of the time you'll be running chip loads in the .002″ to .005″ range and often less than that. In aluminum you can get away with chip loads in the .015″ range if your setup and cutter are rigid.

29. *Use "Surface Feet per Minute" to set spindle speeds.*

This is the number that can help you set efficient spindle speeds. Your goal is to use the highest RPM possible consistent with reasonable tool life so that you can maximize feed rates while maintaining reasonable chip loads.

The formula for determining surface feet per minute is as follows:

$$\text{Surface Feet per Minute (SFM)} = (\text{RPM} \times \text{Cutter Dia.}) \times .262$$

SFM numbers carry more significance when machining tougher materials. In the soft materials like aluminum and brass you can darn near get away with anything. Bear in mind that just because charts recommend using high SFM numbers for soft materials such as aluminum, it is not necessary to do so and may even be detrimental in terms of promoting chatter.

I use the following chart to reference SFM rates for various materials. It is a simple chart but one I consider adequate. It'll get you in the ballpark for most materials. You can, of course, use more detailed charts elsewhere if you prefer. The latest edition of *Machinery's Handbook* has about fifty pages of data on this subject.

As a rule of thumb when using carbide end mills, you can use about three times the SFM and one third the chip load as the values indicated in the chart. When using carbide inserts you can use about three times the SFM and approximately the same chip loads as the numbers indicated below.

Material to be Cut	Feed Per Tooth 1/4" HSS End Mill	Feed Per Tooth 1" HSS End Mill	SFM (HSS)	
			Roughing	Finishing
Aluminum	.003	.01	600	800
Bronze, Medium	.003	.007	250	300
Bronze, Hard	.002	.005	150	200
Cast Iron, Soft	.003	.008	60	80
Cast Iron, Hard	.002	.005	50	70
Steel, Low Carbon	.001	.004	75	90
Steel, 4140	.0005	.003	50	70
Steel, 4340	.0003	.002	50	70
Steel, Stainless, 304	.001	.004	55	75
Steel, Stainless, 17-4 PH	.0005	.003	40	50
Titanium alloy	.001	.004	20	30
Inconel, Hard	.0002	.003	10	20

30. *Use the "Cutting time" formula for determining cutting time.*

The formula is as follows:

Cutting time = Cubic inches of stock to remove /
(Feed Rate x Depth of Cut x Cutter Dia. x Stepover Percentage)

The "cutting time" number is the only number dependent on depth of cut.

Generally speaking when taking a full width cut using a strong cutter in a light machine, I don't like to cut more than .050″ deep and often less in the tougher materials like stainless and from .050″ to .1″ deep in softer materials like aluminum.

Some machinists may consider those roughing cuts to be on the shallow side but I prefer bumping up feed rate rather than burying a cutter to remove material quickly.

There are a few reasons why I think shallow cuts are better than deep cuts.

First, deep cuts put more pressure on the workpiece and cutter which often causes unstable cutting conditions such as cutter and table deflection. Flexing and bouncing cutters always generate more friction and wear faster then a cutter cutting under rigid, stable conditions. Also, when cutting with small diameter cutters you simply can't bury them too deeply or they'll break.

Second, a shallow cut will allow you to increase feed rates to a point that will actually cut the feature faster than if you were to bury the cutter deeper and use a slower feed rate.

Third, chips generated with a shallow cut are able to flow away from the cutting area more easily than larger chips. Chips that don't flow away from the cutting area are destined to be recut which wears cutters faster.

Fourth, with the faster feed rates that shallow cuts allow, much of the heat generated comes out in the chip. This is more noticeable on conventional machines. Your part will stay nice and cool by taking shallow cuts at high feed rates.

Another reason I prefer shallow cuts over deep cuts is because faster feed rates are more impressive. Managers love to see things moving around in a hurry. Even if overall cutting time is the same, the faster feed rates possible with shallow cuts will get you more "points" than deeper cuts with slower feeds.

By using the above "cutting time formula" you can see that the following relationship exists: From any given feed rate and depth of cut, doubling the feed rate and halving that depth of cut will give you the same overall cutting time.

With small cutters in the ⅛″ diameter range and less, you may end up taking no more than a few thousandths of an inch at a time in depth as you run around a contour or feature. I prefer using carbide when using small cutters because of its rigidity.

31. *Use lighter chip loads with helical carbide end mills.*

Carbide end mills simply can't take as much abuse as high speed steel or cobalt end mills. Vibration, chatter, high frequency chatter, chip packing and chip re-cutting are the enemies of carbide and will promote rapid edge break down. Carbide end mills shine mostly in their ability to cut hard and/or abrasive materials without wearing too quickly. They are also more rigid than HSS end mills. Carbide end mills work best as finishing tools on rigid setups, taking light cuts with high surface feet per minute numbers. As a rule of thumb, with solid carbide end mills use about one third the chip load and three times the surface feet per minute settings as you would with an equivalent high speed steel or cobalt end mill.

Carbide inserts by virtue of their economy and compactness can take larger chip loads when cutting conditions are rigid. Insert cutters or cobalt corncob cutters are often a better choice for roughing than solid carbide end mills because they can take more abuse.

32. *In a CNC mill, increase rigid tapping speed the easy way.*

Most of us know that matching spindle RPM and feed is critically important for successful rigid tapping. One way to edit and reduce rigid tapping cycle times without going back to the computer is to apply this simple rule: If you double the spindle RPM you have to double the feed rate of the tap.

For example if the initial spindle speed input in the CAM software was one hundred RPM for tapping a 1/4–20 thread then the computer would output the feed rate in the program to be five inches per minute. Using the above rule to edit the program at the machine, if you double the RPM to two hundred RPM then you have to double the feed rate to ten inches per minute.

You can also use the following formula for determining milling machine rigid tapping speeds and feeds: Feed rate (in/min) = RPM/Threads per inch

In a CNC lathe since feed is set in inches per revolution, once you have the proper rigid tapping relationship, changing the spindle speed does not alter the relationship.

The formula for determining feed per revolution for rigid tapping in a lathe is as follows:

Feed per revolution = 1/Number of threads per inch.

33. *Make a note of what happens when you input values into the "tool wear" registers of CNC machines.*

- In a CNC mill, insert a negative value in the corresponding tool wear register to cut more off inside and outside profiles or diameters.

- In a CNC lathe, insert a positive value into the corresponding wear register to cut more off an inside diameter and a negative value to cut more off an outside diameter.

Note that you can make diameter adjustments within the program itself at the cost of "loosing" or changing nominal size values. I prefer making diameter adjustments by inserting values in the wear registers so that I can refer back to the program for nominal sizes. Be sure to clear wear registers after completing a job to avoid dimensional errors on upcoming jobs.

34. *Quick change tap holders work well in CNC machines. (See Fig. 15-8)*

It is better to use quick change tap holders (rather than collets) for rigid tapping because they provide a little free play with the tap which helps reduce tap failure. When taps do break or get dull they can be quickly changed in these holders.

FIGURE 15–8 Quick change tap holders allow dull and broken taps to be changed quickly.

35. *In a CNC mill, avoid storing tools in the turret that extend below the top of the workpiece.*

You have to be especially careful if you store long reamers and drill bits in the turret. To avoid the risk of indexing such tools into the workpiece during a tool change, install them in the spindle manually.

36. *Use a slitting saw blade to clamp down thin, round parts. (See Fig. 15-9)*

Round parts that need profiling all the way around usually require some kind of special clamping arrangement. You can't use strap clamps coming in from the outside because they would interfere with the tool path. When the part has a hole in the center, a circular clamp of the appropriate diameter can be used to apply pressure near the profile. I've found that slitting saws work great for this purpose. Slitting saws that vary is diameter and thicknesses are available in most shops. Grab one that fits best and use it to clamp down round parts that need edge profiling.

FIGURE 15–9 Slitting saws make great clamps for holding and machining circular parts.

37. *When pocketing or profiling in a* CNC *mill, use an end mill that is slightly smaller than the corner fillets of the part.*

    In a conventional mill, if you need to cut a pocket or profile with a given corner radius, then you're oblidged to use an "on size" end mill to create the correct radius. In a CNC machine, when using CAM software to generate program code, you can choose an end mill whose radius is slightly smaller than the corner radius of the part. The advantage to choosing a smaller end mill is that the end mill will cut with less pressure and have less tendency to chatter as it works its way around the fillet.

38. *What is the difference between contouring and profiling as it relates to machining in a milling machine?*

    Profiling is a term used to describe 2D milling whereas contouring is a term that can be used to describe both 2D and 3D milling.

chapter **16** # Shop Math Wizardry

With the advent of CAD systems, fewer and fewer people are solving shop math problems the old fashioned way. Nevertheless, hardly a day goes by that a machinist doesn't make a few calculations of some sort.

With the vast majority of shop math being simple addition and subtraction, many machinists can get by with few math skills. Sooner or later, as a machinist you'll be faced with a problem that elementary math won't solve. If you are able to do trigonometry; find the circumference of a circle and use Pythagorus's theorem then you'll be able to solve most shop math problems.

Like many skills with learning curves, practice makes perfect. Math is no exception. In this chapter, I'd like to present what I think are some interesting and relevant math problems for readers to practice on. Some of these problems are easily solved using a CAD system. Is that cheating? I suppose not. But if you enjoy a challenge then try working them long hand to test and improve your math and reasoning skills.

These are not easy problems for many people so don't get discouraged if you can't get them. Maybe let your co-workers have a stab at them. At the very least you may end up learning something. The answers to these problems are shown in Appendix 2.

1. *How much material must be ground off each wear plate in the diagram below so the cavity blocks can sit flush and centered in the mold base?*

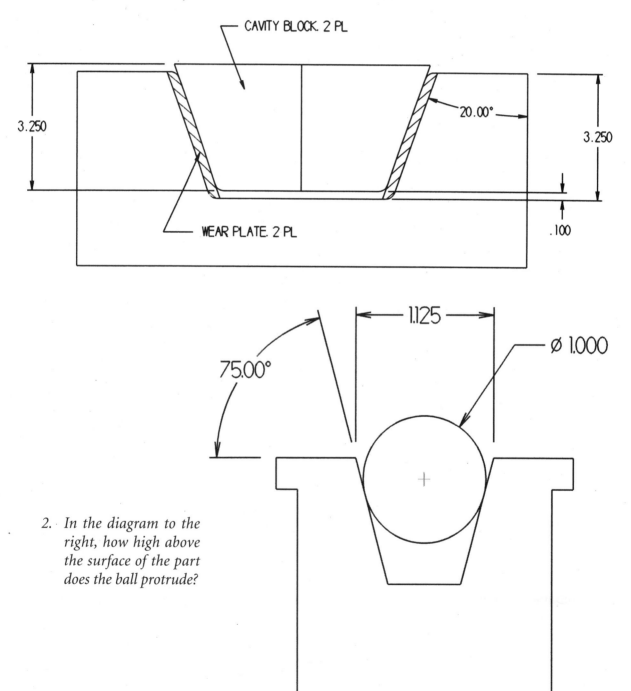

CAVITY BLOCK. 2 PL

20.00°

3.250

3.250

WEAR PLATE. 2 PL

.100

1.125

Ø 1.000

75.00°

2. *In the diagram to the right, how high above the surface of the part does the ball protrude?*

3. *How many inches per minute would you feed an M6 X 1.25 tap rotating at 300 RPM for rigid tapping?*

4. *A band saw has two drive wheels each 2 ft. in diameter. The blade tension adjuster can change the center to center distance of the wheels from 2' 10" to 3' 2". What are the maximum and minimum blade lengths that can be used to make blades for this saw?*

5. *A square based pyramid is made with four walls that angle in at forty-five degrees. What are the angles of the corners of the pyramid with respect to the base?*

6. *A string is tied around the circumference of the earth. (assume the string has no thickness) The string is then cut and another piece of string is added or spliced into the original piece so that a one-inch gap can now be made between the circumference of the earth and the string all the way around. How much string must be added? Do the same problem with any other ball.*

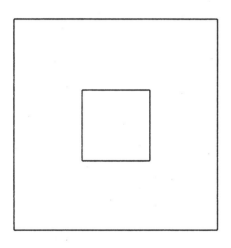

7. *This problem came to us from our injection molding department: How many pounds of nylon would you have to add to a hundred pound mixture of plastic consisting of 80% glass and 20% nylon to form a new mixture consisting of 60% glass and 40% nylon?*

8. *(See diagram to the left.) This isn't a math problem but I like it anyway: Two views of a part are shown in the drawing. Draw a third view that has no hidden lines.*

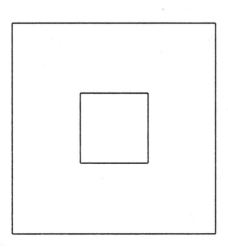

?
.

9. *This problem initially looks simple. It's not, unless you use a computer:*
    A ten-foot ladder is leaned up against a wall. A two-foot cubic box is placed under the
    ladder and up against the wall. The angle of the ladder is adjusted so that it is touching
    the floor, the wall, and the corner of the box. How high up the wall does the ladder
    touch?

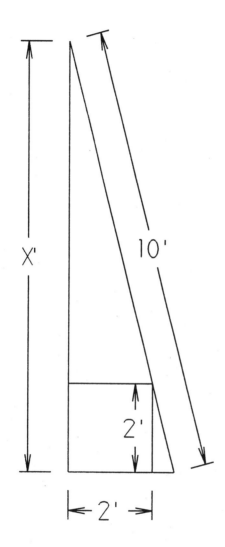

chapter **17** **Odds and Ends**

Engineering Terms You May Hear

1. *What is tensile strength?*

    This is a commonly used term when talking about the strength of materials. It is the load, usually stated in PSI necessary to break a piece of material by pulling it apart.

2. *What is stress?*

    Stress is the load or pressure applied to something. Stresses relevant to machining are those applied to cutting tools, fixtures and parts while they are being machined.

3. *What is strain?*

Strain is material movement. For example, when a boring bar or end mill flexes under pressure or stress, it is said to be strained.

4. *What does yield point refer to?*

If you clamp a piece of bar stock in a bench vise and start pulling on the end of it, in the beginning you'll notice that the more you bend the bar the more force it will take to continue the bending. When you reach an angle where it no longer takes extra force but just steady force to keep the bend going that means the material is yielding.

Yield points can be shown graphically as the point of departure from a straight line on a stress/strain graph.

Things I'd Like to See

5. *I'd like to see more information in* Machinery's Handbook *about "galling".*

The word isn't even listed in the index. Based on my experience it is an important issue in regards to keeping tools and machinery running. I would like to know what materials, hardnesses and coatings work best and which ones work worst in conjunction with each other to avoid galling problems.

6. *Milling machine vise with no slot in the center.*

I'd like to have a high quality milling machine vise with no slot in the center. The slot in the center of the common Angle Lock® type vises makes it impossible to set small parts in the middle of the vise without using parallels.

7. *Commercially available vise jaws with precisely square ends with 1/4–20 tapped holes in each end that could be used to bolt on a simple vise stop.*

8. *Tall commercially available vise jaws. (See Fig. 17-1)*

Extra tall, sturdy vise jaws come in handy once in a while.

9. *A set of industrial quality drill bits with shanks ground to common fractional collet sizes.*

10. *Bridgeport® milling machine with a slower spindle speed in high gear.*

I'd like to be able to go as low as 300 RPM in high gear instead of the 500 RPM currently available.

FIGURE 17–1 Tall vise jaws come in handy once in awhile.

Worthless Advice?

The following are some marginally valuable tidbits of advice I've received from well mean-
ing mentors.

11. *"Look in the chip pan for answers to your machining problems."*

     Chip form and color can be indicators of proper machining parameters; nevertheless
     perfect chips don't necessarily mean perfect parts.

12. *"Your part is in there somewhere."*

     One guy graciously informed me as I was preparing a piece of stock that my finished
     part was somewhere within that piece of stock. DUH

Processing Odds and Ends

13.   What does "Class" and "Type" refer to when anodizing aluminum?

Class 1 anodizing is non-dyed. Class 2 anodizing is dyed. The color is specified in the contract. Type I anodizing produces a coating of between .0002″ and .0007″. Type II anodizing produces a coating of between .0007″ and .001″.

14.   What is hard anodizing?

Hard anodizing is essentially the same as anodizing except that the process creates a thicker coating in the same amount of time. Unless called out in the contract a typical hard anodized coating would be .002″ thick including penetration.

15.   What is passivating?

It is a common process that uses a nitric acid dip to treat stainless steel. The process does not remove any parent material but is used to dissolve iron particles that may have imbedded in parts during machining or handling. It also has the benefit of revealing parts that were made from the wrong material by severely attacking them.

16.   What is black oxiding?

It is a process that creates a mostly decorative black coating on ferrous metals. The coating offers limited corrosion protection.

17.   What is electropolishing?

It is a process used to smooth and brighten stainless steel. This process removes a slight amount of parent material.

18.   What is hard chrome plating?

It is a process that can be used to build up surfaces of undersize or worn out steel parts with a tough, wear resistant chromium plating. Piston rings are covered with hard chrome. The average life of a hard chrome plated piston ring is about five times that of an unplated ring. The hard chrome can be deposited in varying thicknesses. Hard chrome differs from decorative chrome in that decorative chrome is very thin and is applied over an undercoating of nickel or nickel and copper.

Heat Treating and Metallurgical Stuff

19. *What are austenitic steels?*

They are steels that cannot be hardened by heat treating.

20. *What are martensitic steels?*

They are steels that can be hardened by heat treating.

21. *What does "critical temperature" refer to?*

The elevated temperature at which the dissolving of all alloys and carbon occurs is called the critical temperature. Since the metal is still in a solid state it is said to be in solid solution.

22. *What does precipitation hardening refer to?*

Some materials can be hardened by heating them below critical temperature and letting them cool slowly at room temperature. As these materials cool, nitrides precipitate out of the solid solution and act as "keys" to strengthen the material. 17-4PH and 15-5PH steels are commonly heat treated this way.

23. *What is nitriding?*

It is a case hardening operation carried out in an atmosphere of ammonia at the relatively low temperature of 1000° F. Not all steels can be case hardened this way. Steels containing chromium and molybdenum should be used. Since the process doesn't require quenching, little distortion occurs. Ejector pins have nitrided surfaces.

24. *Is there a difference between drawing steel and tempering it?*

They are the same thing and refer to the process of heating up hardened material below critical temperature and letting it cool slowly to reduce its hardness or brittleness and increase its toughness.

25. *What is annealing.*

It is the process of softening a hardened piece of steel by heating it above its critical temperature and letting it cool slowly.

Welding and Brazing Odds and Ends

26. *Set oxyacetylene torch regulators and needle valves based on flame appearance.*

    The following are the basic procedures for setting up a shop torch.

    a. Close both regulator valves by turning the regulator screws counterclockwise until they turn freely. Also close the needle valves on the torch.

    b. Open the main acetylene cylinder valve 1/4 to 1/2 turn. Open the main oxygen cylinder valve all the way.

    c. Open the acetylene needle valve on the torch one full turn.

    d. Turn the acetylene regulator clockwise until gas starts coming from the tip of the torch. Light the gas.

    e. Adjust the acetylene regulator until there is a 1/4" gap between the flame and the tip. This is the correct pressure regardless of tip size.

    f. Open the oxygen needle valve on the torch one full turn. Open the oxygen regulator by screwing the handle clockwise until the flame starts to change appearance.

    g. Adjust the oxygen regulator until the acetylene feather just disappears into the inner cone. This is a neutral flame.

    The above procedures establish proper pressures in the equipment which eliminates the possibility of backing gas from one hose to another. Once proper pressures have been established with the regulators, minor flame adjustments can be made with the torch.

27. *Use a slightly carburizing flame when brazing and silver soldering.*

    • Oxidizing flame (excess oxygen):
    Small blue inner cone at the base of the flame.

    • Carburizing flame (excess acetylene):
    "Feather" extending slightly beyond the inner cone.

    • Neutral flame (equal volumes of oxygen and acetelyne):
    "Feather" just disappears into the inner cone.

28. *Use a silver alloy with nickel and cadmium in it for general purpose shop work.*

    Silver solders containing cadmium and nickel are well suited for silver soldering carbide, tool steel, stainless steel and other materials. The nickel provides corrosion resistance, the cadmium lowers the melting point and makes the alloy flow better.

29. *The higher the silver content of silver solder the less it will flow and the stronger it will be.*

30. *When silver soldering, cover the surface of the part liberally with flux to keep surfaces from oxidizing.*

    Keep the torch moving when silver soldering so you don't burn through the flux in one area and create surface oxidation. Dip or rinse the part in hot water when you are done to dissolve the solidified flux.

31. *Why do torches sometimes go out with a loud bang?*

    Holding the tip of a torch too close to the work or starving the flame of acetylene will cause a torch to go out with a loud bang. Avoid letting the acetylene tank pressure get any lower than about 50 PSI. Make sure there are no leaks in any of the connections.

32. *Use a Mapp® gas in combination with a self lighting torch head to avoid having to set up an oxyacetylene torch. (See Fig. 17-2)*

    Mapp gas torches work fine for the majority of the silver soldering jobs you'll encounter in a machine shop.

FIGURE 17–2

Mapp® gas torches get hot enough to do silver soldering.

33. *The following is a list of approximate temperatures reached by various torches and fuels:*

    - Propane torches reach about 2500° F.

    - Mapp gas torches reach about 3600° F.

    - Oxyacetylene torches reach about 6300° F.

    - The surface of the sun reaches about 10,800° F.

    - An electric arc from an arc welder reaches about 19,800° F.

    As a point of reference:

    - Steel starts to turn red at about 1000º F.

    - Silver solder melts in a range from about 1100º to 1300º F.

- 6061-T6 aluminum melts at about 1150º F.

- Low carbon steel melts at about 2700º F.

- Tungsten melts at about 6170º F.

34. *What is the difference between Tig and Mig welding?*

TIG stands for "Tungsten Inert Gas" and is a welding process that uses a non-consumable electrode. The process is well suited for welding thin metal.

MIG stands for "Metal Inert Gas" and is a welding process that uses a continually fed consumable wire as the electrode.

35. *6061 aluminum welds better than 7075 or 2024 aluminum.*

36. *What does the "T" stand for in 6061-T6 aluminum.*

The "T" designation following the type of aluminum is the heat treat condition or hardness of the aluminum. The higher the number the harder the material. The number "511" which sometimes follows the hardness designation refers to the fact that the aluminum was extruded rather than cast. (i.e. 6061-T6511)

37. *Before welding a bushing into a part, it's good practice to install a close fitting plug into the ID of the bushing so it doesn't distort or collapse during welding.*

Yes We Can Save It

38. *Use penetrating oil and heat from a torch to loosen frozen parts.*

39. *Use a hammer and aluminum punch to rap the head of a tight bolt.*

The first thing you should do when trying to remove a tight bolt is rap the head of the bolt with an aluminum rod and hammer. The shock of the impact will help loosen the bolt.

40. *Use a prick punch in combination with a wrench to remove tight bolts.*

Pressure plus shock is better than either pressure or shock for loosening tight bolts. A sharp center punch and hammer can be used to apply a tremendous amount of impact torque to the head of a bolt. The impact torque in combination with steady wrench torque will loosen just about any tight bolt. Either that or it will break the head off the bolt.

41. *Use left handed drill bits to drill out broken bolts. (See Fig. 17-3)*

If the head of a bolt breaks off then you're probably going to have to drill out the remaining material. Left hand drill bits work great for this purpose. Begin by drilling a pilot hole through the remaining bolt material about one third the diameter of the final drill diameter you choose. The pilot hole removes a little strength from the bolt and allows you to feed the left hand bit aggressively into the bolt creating a lot of counterclockwise torque. Often the torque created by the left hand bit will wind out the remaining portion of the bolt. The final diameter of the bit you choose should be somewhat under the minor diameter of the tapped hole so threads don't get damaged.

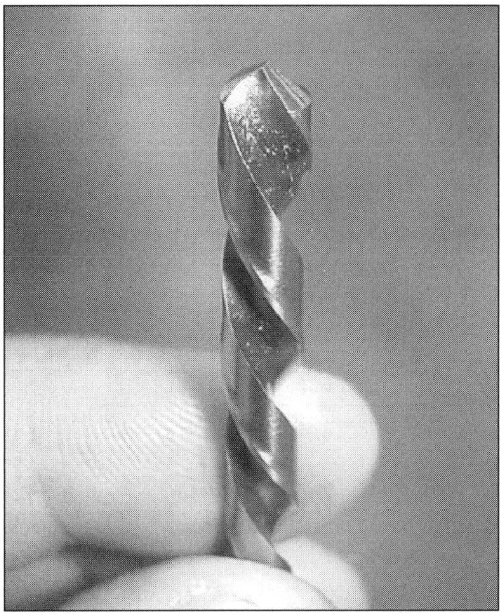

FIGURE 17–3 Left handed drill bits work well for drilling out broken bolts.

42. *Use Keenserts® to repair damaged internal threads. (See Fig. 17-4)*

I prefer using Keenserts® to rebuild damaged or worn out internal threads. They are rugged, stay in place well and can be installed with common shop tools.

43. *Use a six sided ball nose carbide cutter to remove broken taps. (See chapter 8 for details)*

44. *Use compressed air and vise grips to remove broken drill bits.*

Broken drill bits are usually not too difficult to remove since they are not

FIGURE 17–4 Keenserts® can be used to rebuild stripped and worn out threaded holes.

locked in place like a tap. Removing chips and debris from around the broken bit with compressed air will usually free it sufficiently so that it can be extracted.

45. *Correct a hole that has been bored slightly off location.*

One easy way to correct a hole that has been bored slightly off location is to bore a bigger hole in the right location and install a sleeve.

Things That Can Be Irritating

47. *Tap extractors because they rarely work.*

48. *Missing tools from sets such as gauge pins and reamers.*

An effective shop policy to avoid loosing tools from sets is to have shop personnel take the full set back to their work station. That way when a person is finished using a tool they are less likely to inadvertently put the tool in their toolbox.

49. *Brittle taps.*

Some manufacturers produce taps that are so brittle as to be virtually worthless.

50. *Lathe chucks that can't be adjusted to run concentric.*

51. *Long stringy chips.*

52. *Machinists that don't disassemble and put away their tie-down clamp hardware.*

53. *People that return borrowed tools with chips and oil on them and people that don't return them at all.*

54. *Lousy edge finders that don't jump abruptly or far enough.*

55. *Milling machines with sticky quills.*

56. *Digital readouts that skip.*

57. *Smoky cutting oil.*

58. *Lousy drawings.*

59. *Drill presses, arbor presses and other small pieces of equipment that are not bolted to the floor.*

60. *Insert type cutters and tool holders that can't be rebuilt with standard hardware.*

61. *And last but not least; taking over a job that someone else started.*

chapter 18

Tell Me Something I Don't Know

1. *According to the bureau of labor statistics for the year 2001, the median hourly wage for tool and die makers in the U.S. was $20.01/hr.*

   The top ten percent were making $29.49/hr. Michigan and Washington had the highest median wage at $24.29/hr. and $23.67 respectively. Mississippi and West Virginia had the lowest median wage at $13.76/hr. and $12.48/hr.

2. *According to my high school shop teacher, a drill press causes more injuries than any other type of metal working machine. Lathes and disk sanders come in a close second.*

3. Wood's metal has the lowest melting point of any solid metal at room temperature. It melts at 158° F. It consists of about 12% Cadmium, 12% Tin, 24%, Lead and 50% Bismuth.

4. The average automobile consists of about a hundred and fifty different alloys of metal.

5. An average man working hard can output about one quarter horsepower.

6. Loctite ® thread and bushing lockers do not dry like glue, they set up in a lack of oxygen.

7. A one hundred degree ºF rise in temperature will expand steel approximately .0006" per inch and aluminum about twice that amount.

8. Something that is 140º to 150º F is difficult to touch or hold for an extended length of time.

9. The "C" in 5C collet stands for "Cataract" which is the name of the waterfalls that could be seen from the windows of the original Hardinge factory in Chicago, Ill.

10. The "RS" in RS232 stands for "Recommended Standard."

11. The "HTTP" you see at the beginning of a web address or URL stands for "Hypertext Transfer Protocol."

12. "URL" stands for "Uniform Resource Locator."

13. A U.S. five cent piece consists of 75% copper and 25% nickel.

14. The plating on U.S. coinage consists of 75% copper and 25% nickel.

15. You can soft solder aluminum if you break the oxide layer by scratching the surface of the aluminum with a hard metal object through the molten solder.

16. Most rifle barrels are made of either 4140, 4150 ordinance steel or 416R stainless steel heat treated to about 30 RC. The average shot gun barrel develops 1600 pounds per square inch of pressure when fired.

17. *Assuming equivalent gooseneck arrangements and indicator sweep areas, a more accurate milling machine tram can be made with the quill extended.*

18. *How are metal Slinkies® made?*

Some lathe chips look surprisingly a lot like Slinkies. The observation may lead some people to believe that Slinkies are nothing more than controlled lathe chips. Not true. Real Slinkies are made from round stainless steel wire which is pressed flat and coiled in special machines.

19. *How many lbs. of clamping force does a 1/2 – 13 bolt tightened to 40 foot pounds exert?*

Torque values applied to a bolt or nut only give an approximation of the resulting clamping pressure. Roughly 40% of the torque applied to a bolt is used to overcome thread friction. Another 40% or so is used to overcome bolt face friction. The remaining 20% of the applied torque actually "stretches" the bolt and applies clamping pressure.

Most clamping and bolting that goes on in a machine shop is done with dry threads. A rough approximation for the clamping pressure applied by a 1/2 – 13 bolt torqued to 40 foot lbs in a dry condition is about 1000 lbs.

20. *What causes metals to work harden?*

You can start an interesting debate with this question. Some people say work hardening happens as a result of the heat generated while machining which on a small scale heat-treats the material into a harder condition.

Other people say it happens as a result of material being pushed, upset, deformed and forced into plastic movement as it is being machined which causes changes in the grain structure and work hardening. I tend to agree with the "material movement" people. You can minimize work hardening by using sharp cutters and not dwelling cutters in the workpiece. Most steel alloys are susceptible to work hardening.

21. *What is the difference between a gauge and a gage?*

The only difference is the spelling. "Gauge" is the more common spelling.

22. *What is a cone shaped edge finder used for? (See Fig. 18-1)*

Opinions vary. Some say they're used to find the center of a hole. Others say they're used to line up prick punch marks. The most practical use I've found for one is to find the center of a narrow slot. You do that by inserting the tip of the cone into a slot at some arbitrary depth and touching off both sides until the tip jumps noting the total travel of the table with the digital readout. Divide that number by two to position the spindle over the center of the slot.

23. *What differences are there in granite surface plates?*

Black granite surface plates can carry more weight without distorting than plates high in quartz. They, however, wear faster than high quartz plates. Black granite is less prone to water absorption and is generally more stable than the high quartz variety.

FIGURE 18–1 If you'd like to start an interesting debate ask a machinist or engineer what a cone shaped edge finder is used for.

24. *Nickel base alloys such as Inconel have some of the lowest machinability ratings of all alloys.*

To turn a six inch diameter part at 30 SFM which is the average recommended SFM rate for cutting tough nickel base alloys with carbide, the spindle would have to be turning at 19 RPM. Most small engine lathes won't even turn that slow.

25. *What are "refractory metals"?*

"Refractory metals" refer to metals with high melting points. A refractory material is one that resists change of shape, weight, or physical properties at high temperatures. Some of the more common refractory metals are: Tungsten, Molybdenum, Tantalum, Niobium, Chromium and Vanadium. As a point of reference, Tungsten melts at about 6170° F. Low carbon steels melt at about 2700° F.

26. *What is dental amalgam?*

It is a mixture of mercury, silver, tin and copper. The compound consists of about 45-50% mercury. The mercury binds the alloys together to form a strong, durable filling.

27. *How much does it cost to run a small milling machine for eight hours?*

Less than a cup of coffee. Assuming electricity costs seven cents per KWH. The electricity to run a two HP three phase milling machine under a 50% load for eight hours would cost about forty three cents. A twenty horsepower machine using the same parameters would cost about ten times that much or $4.30.

FIGURE 18–2 The angle on this part was created by cutting material away in small tracks…a process called "kellering".

28. *Surface finish measurements are defined as the average deviation (Ra) from the mean line of the surface texture commonly expressed in microinches. As a point of reference, the surface textures of the following items were measured.*

- Pane of glass: 2
- Steel block ground with a 60 grit wheel: 6–8
- Steel block ground with an 80 grit wheel: 6–8
- Steel block ground with a 46 grit wheel: 15–18
- Piece of copy paper: 14–20
- New dollar bill: 49–57
- Piece of masking tape: 97–114
- 500 grit wet/dry sandpaper: 115–120
- 33 RPM vinyl record: 240–250
- 45 RPM vinyl record: 327–340

29. *What is kellering? (See Fig. 18-2)*

Kellering is the process of generating a smooth surface by cutting material away in small steps or tracks. The process is well adapted for cutting lofted surfaces. It is usu-

FIGURE 18–3 A file is friction sawed in two using a worn out blade in a vertical bandsaw.

ally done in a CNC mill but can also be done in a conventional machine. Ball nose end mills are commonly used for kellering since they provide the best surface finish for a given end mill diameter and step over distance. Bull nose and flat bottom end mills can also be used. The angled feature in the picture was kellered in a CNC mill with a flat bottom end mill to produce a sharp corner at the intersection of the base and the angle.

30. *Can you cut a file with a band saw? (See Fig. 18-3)*

You can. Install an old worn out blade in the saw and run it as fast as it will go. The blade will friction cut its way through the file with little difficulty.

31. *Can you do a job in a "jiffy"?*

Probably not. A "jiffy" is an actual unit of time equal to 1/100 of a second.

FIGURE 18–4 A vise can be lined up relatively close by eyeballing a bar across the edge of a machine table.

32. *Eyeball the squaring of a milling machine vise within .002" everytime? (See Fig. 18-4)*

This suggestion may not have much practical value but it may win you a bet. If you clamp a long rigid straight edge in a milling machine vise, you can eyeball the straight edge over the edge of the table and bump in the vise. You better practice a few times before you take a bet.

Appendix 1

Project Drawings

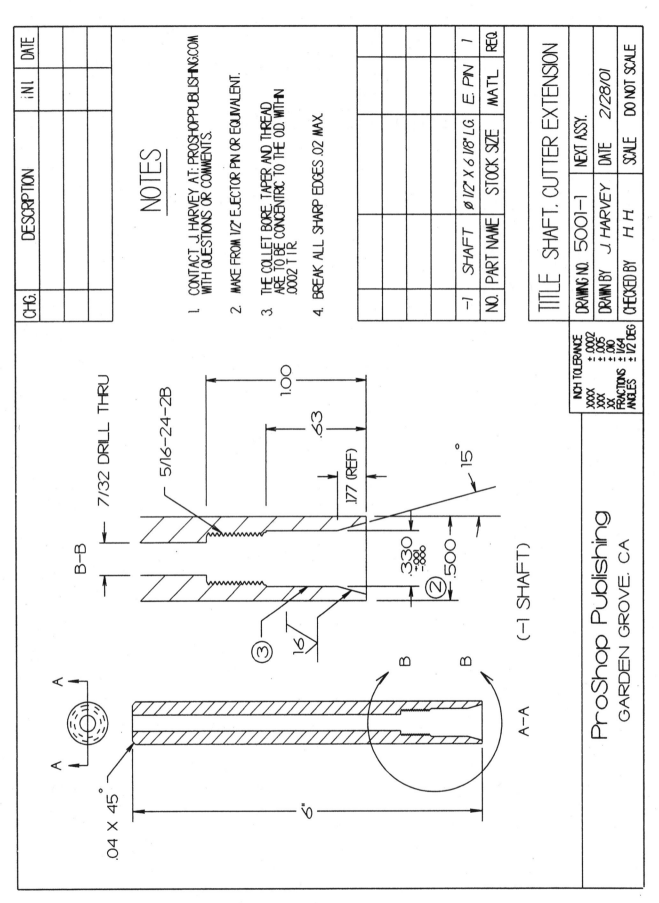

NOTES

1. CONTACT J HARVEY AT: PROSHOPPUBLISHING.COM WITH QUESTIONS OR COMMENTS.

2. MAKE FROM 1/2 EJECTOR PIN OR EQUIVALENT.

3. THE COLLET BORE, TAPER AND THREAD ARE TO BE CONCENTRIC TO THE O.D. WITHIN .0002 T.I.R.

4. BREAK ALL SHARP EDGES .02 MAX.

CHG.	DESCRIPTION	INT	DATE

NO.	PART NAME	STOCK SIZE	MATL	REQ.
-1	SHAFT	Ø 1/2" X 6 1/8" LG.	E. PIN	1

TITLE SHAFT, CUTTER EXTENSION

NEXT ASSY.		
DRAWING NO. 5001-1		
DRAWN BY J. HARVEY	DATE 2/28/01	
CHECKED BY H.H.	SCALE	DO NOT SCALE

INCH TOLERANCE
.XXXX ± .0002
.XXX ± .005
.XX ± .010
FRACTIONS ± 1/64
ANGLES ± 1/2 DEG

ProShop Publishing
GARDEN GROVE, CA

7/32 DRILL THRU

5/16-24-2B

1.00

.63

.177 (REF)

15°

.330

.500

B-B

.04 X 45°

6"

A-A   (-1 SHAFT)

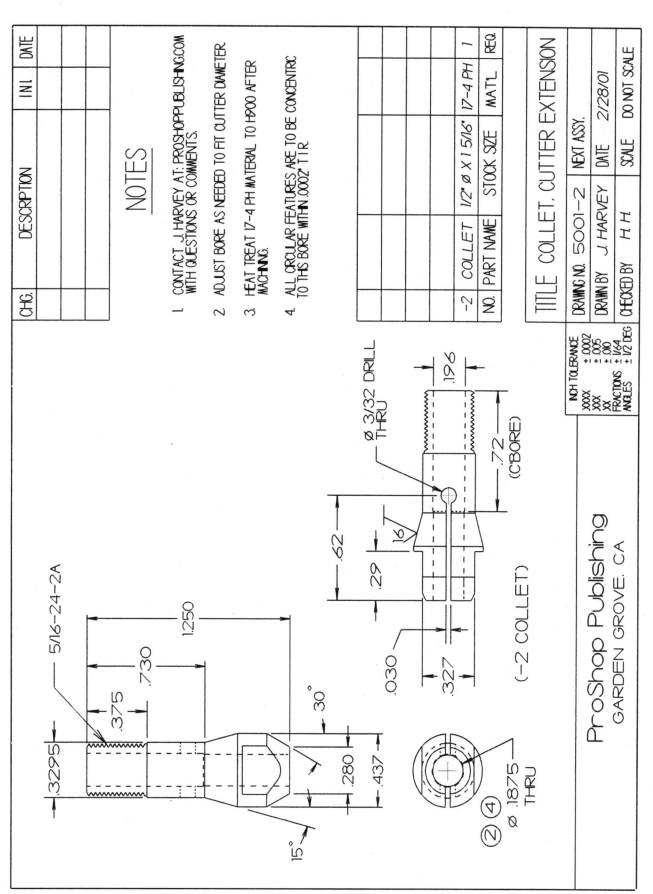

NOTES

1. CONTACT J. HARVEY AT: PROSHOPPUBLISHING.COM WITH QUESTIONS OR COMMENTS.

2. ADJUST BORE AS NEEDED TO FIT CUTTER DIAMETER.

3. HEAT TREAT 17-4 PH MATERIAL TO H900 AFTER MACHINING.

4. ALL CIRCULAR FEATURES ARE TO BE CONCENTRIC TO THIS BORE WITHIN .0002 T.I.R.

CHG.	DESCRIPTION	INI.	DATE

NO.	PART NAME	STOCK SIZE	MAT'L	REQ.
-2	COLLET	1/2" Ø x 1 5/16"	17-4 PH	1

TITLE COLLET, CUTTER EXTENSION

DRAWING NO. 5001-2	NEXT ASSY.
DRAWN BY J. HARVEY	DATE 2/28/01
CHECKED BY H.H.	SCALE DO NOT SCALE

ProShop Publishing
GARDEN GROVE, CA

INCH TOLERANCE	
.XXXX	± .0002
.XXX	± .005
.XX	± .010
FRACTIONS	± 1/64
ANGLES	± 1/2 DEG

Ø 3/32 DRILL THRU

.196

.72 (CBORE)

.62

.29

.16

.030

.327

(-2 COLLET)

5/16-24-2A

1.250

.730

.375

.3295

30°

.280

.437

15°

Ø .1875 THRU

② ④

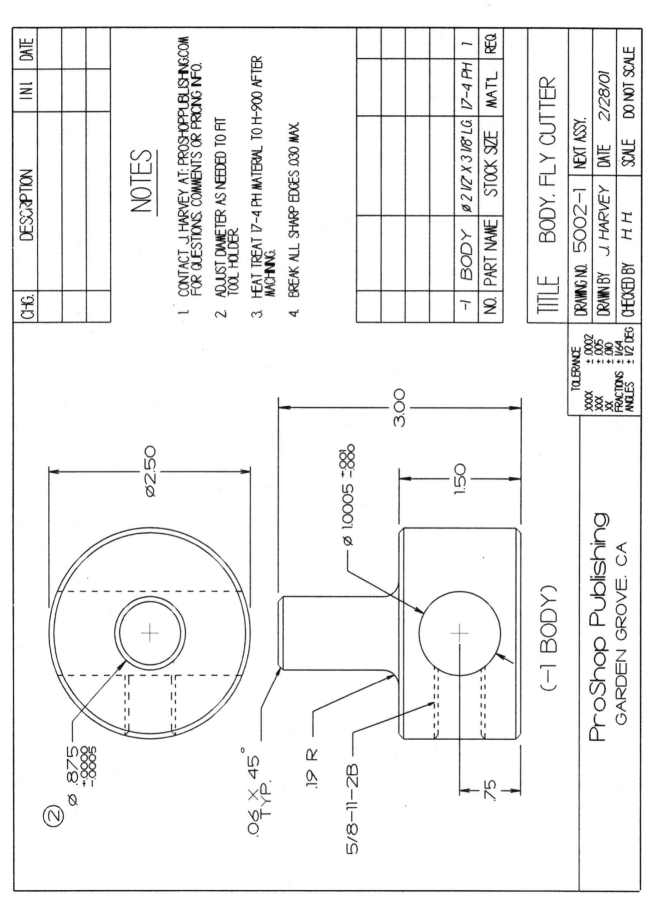

NOTES

1. CONTACT J. HARVEY AT: PROSHOPPUBLISHING.COM FOR QUESTIONS, COMMENTS OR PRICING INFO.

2. ADJUST DIAMETER AS NEEDED TO FIT TOOL HOLDER.

3. HEAT TREAT 17-4 PH MATERIAL TO H-900 AFTER MACHINING.

4. BREAK ALL SHARP EDGES .030 MAX.

CHG.	DESCRIPTION	INT	DATE

NO.	PART NAME	STOCK SIZE	MAT'L	REQ.
-1	BODY	Ø 2 1/2 X 3 1/8" LG.	17-4 PH	1

TITLE   BODY, FLY CUTTER

	NEXT ASSY.
DRAWING NO. 5002-1	
DRAWN BY J. HARVEY	DATE 2/28/01
CHECKED BY H H	SCALE DO NOT SCALE

TOLERANCE
XXXX   ± .0002
XXX   ± .005
XX   ± .010
FRACTIONS   ± 1/64
ANGLES   ± 1/2 DEG

Ø 2.50

Ø .875 +.0000 -.0005

Ø 1.0005 +.001 -.000

.06 X 45° TYP.

.19 R

5/8-11-2B

3.00

1.50

.75

(-1 BODY)

ProShop Publishing
GARDEN GROVE, CA

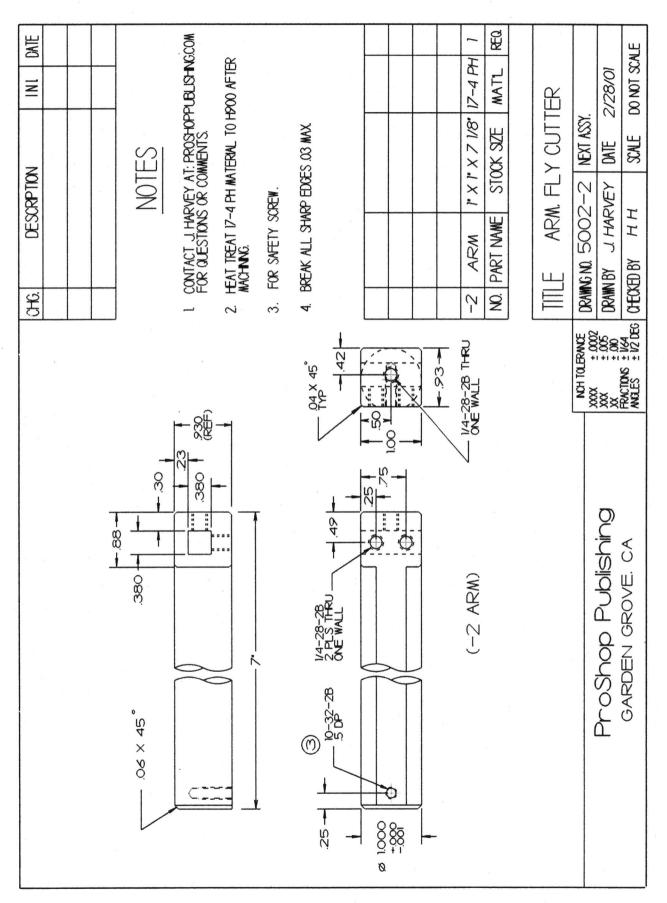

NOTES

1. CONTACT J. HARVEY AT: PROSHOPPUBLISHING.COM FOR QUESTIONS OR COMMENTS.

2. HEAT TREAT 17-4 PH MATERIAL TO H900 AFTER MACHINING.

3. FOR SAFETY SCREW.

4. BREAK ALL SHARP EDGES .03 MAX

CHG.	DESCRIPTION	INT	DATE

NO.	PART NAME	STOCK SIZE	MAT'L	REQ.
-2	ARM	1" X 1" X 7 1/8"	17-4 PH	1

TITLE ARM. FLY CUTTER

DRAWING NO. 5002-2	NEXT ASSY.
DRAWN BY J. HARVEY	DATE 2/28/01
CHECKED BY H.H.	SCALE DO NOT SCALE

INCH TOLERANCE
.XXXX ± .0002
.XXX   ± .005
.XX   ± .010
FRACTIONS ± 1/64
ANGLES ± 1/2 DEG

ProShop Publishing
GARDEN GROVE. CA

(-2 ARM)

.04 X 45° TYP
1/4-28-2B THRU ONE WALL
.06 X 45°
1/4-28-2B 2 PLS THRU ONE WALL
10-32-2B .5 DP
Ø 1.000 +.000 -.001

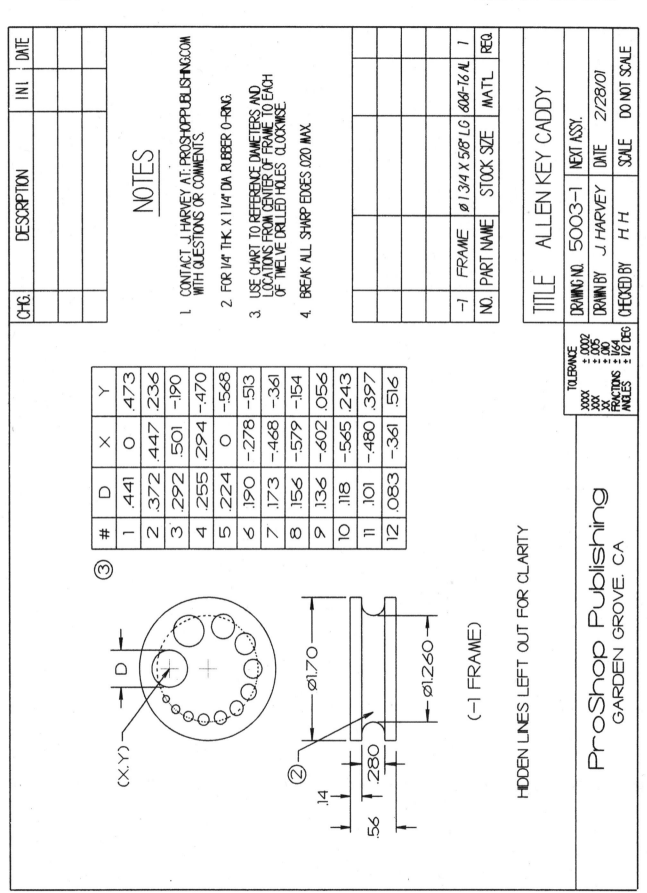

CHG.	DESCRIPTION	INI.	DATE

NOTES

1. CONTACT J. HARVEY AT: PROSHOPPUBLISHING.COM WITH QUESTIONS OR COMMENTS.

2. FOR 1/4" THK. X 1 1/4" DIA RUBBER O-RING.

3. USE CHART TO REFERENCE DIAMETERS AND LOCATIONS FROM CENTER OF FRAME TO EACH OF TWELVE DRILLED HOLES. CLOCKWISE.

4. BREAK ALL SHARP EDGES .020 MAX.

NO.	PART NAME	STOCK SIZE	MAT'L	REQ.
-1	FRAME	ø 1 3/4 x 5/8 LG	6061-T6 AL	1

TITLE	ALLEN KEY CADDY	
DRAWING NO. 5003-1	NEXT ASSY.	
DRAWN BY J. HARVEY	DATE 2/28/01	
CHECKED BY H.H.	SCALE	DO NOT SCALE

TOLERANCE	
.XXXX	± .0002
.XXX	± .005
.XX	± .010
FRACTIONS	± 1/64
ANGLES	± 1/2 DEG

#	D	X	Y
1	.441	0	.473
2	.372	.447	.236
3	.292	.501	-.190
4	.255	.294	-.470
5	.224	0	-.568
6	.190	-.278	-.513
7	.173	-.468	-.361
8	.156	-.579	-.154
9	.136	-.602	.056
10	.118	-.565	.243
11	.101	-.480	.397
12	.083	-.361	.516

(-1 FRAME)

HIDDEN LINES LEFT OUT FOR CLARITY

ProShop Publishing
GARDEN GROVE, CA

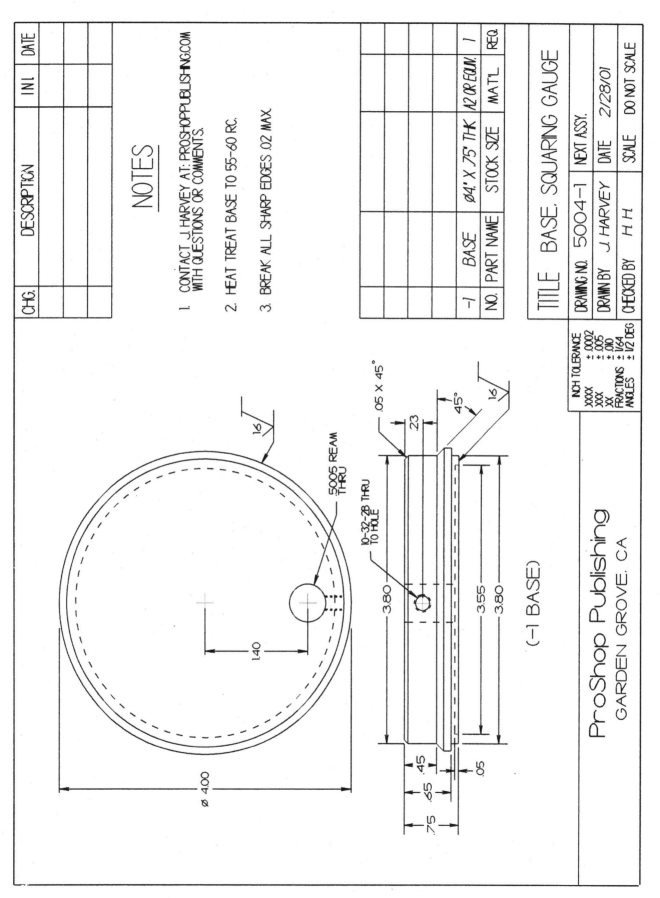

CHG.	DESCRIPTION	INL	DATE

NOTES

1. CONTACT J. HARVEY AT: PROSHOPPUBLISHING.COM WITH QUESTIONS OR COMMENTS.

2. HEAT TREAT BASE TO 55-60 RC.

3. BREAK ALL SHARP EDGES .02 MAX.

NO.	PART NAME	STOCK SIZE	MAT'L	REQ.
-1	BASE	Ø4" X .75" THK	A2 OR EQUIV.	1

TITLE BASE, SQUARING GAUGE

DRAWING NO. 5004-1	NEXT ASSY.
DRAWN BY J. HARVEY	DATE 2/28/01
CHECKED BY H.H.	SCALE DO NOT SCALE

INCH TOLERANCE	
XXXX	± .0002
XXX	± .005
XX	± .010
FRACTIONS	± 1/64
ANGLES	± 1/2 DEG

.5005 REAM THRU

10-32-2B THRU TO HOLE

1.40

Ø 4.00

.05 x 45°

.23

45°

3.80

3.55

3.80

.45

.65

.75

.05

(-1 BASE)

ProShop Publishing
GARDEN GROVE, CA

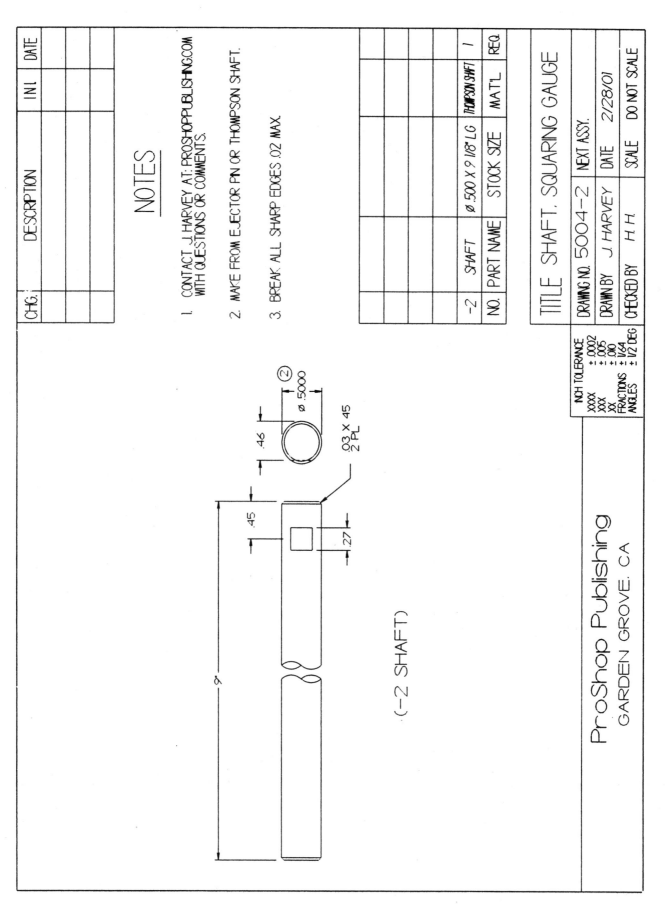

CHG.	DESCRIPTION	INT.	DATE

NOTES

1. CONTACT J. HARVEY AT: PROSHOPPUBLISHING.COM WITH QUESTIONS OR COMMENTS.

2. MAKE FROM EJECTOR PIN OR THOMPSON SHAFT.

3. BREAK ALL SHARP EDGES .02 MAX.

NO.	PART NAME	STOCK SIZE	MAT'L	REQ.
-2	SHAFT	Ø .500 X 9 1/8" LG	THOMPSON SHAFT	1

TITLE   SHAFT, SQUARING GAUGE

DRAWING NO. 5004-2	NEXT ASSY.
DRAWN BY J. HARVEY	DATE 2/28/01
CHECKED BY H.H.	SCALE DO NOT SCALE

INCH TOLERANCE	
.XXXX	± .0002
.XXX	± .005
.XX	± .010
FRACTIONS	± 1/64
ANGLES	± 1/2 DEG

ProShop Publishing
GARDEN GROVE, CA

(-2 SHAFT)

Ø .5000

.46

.45

.27

.03 X 45°
2 PL

9

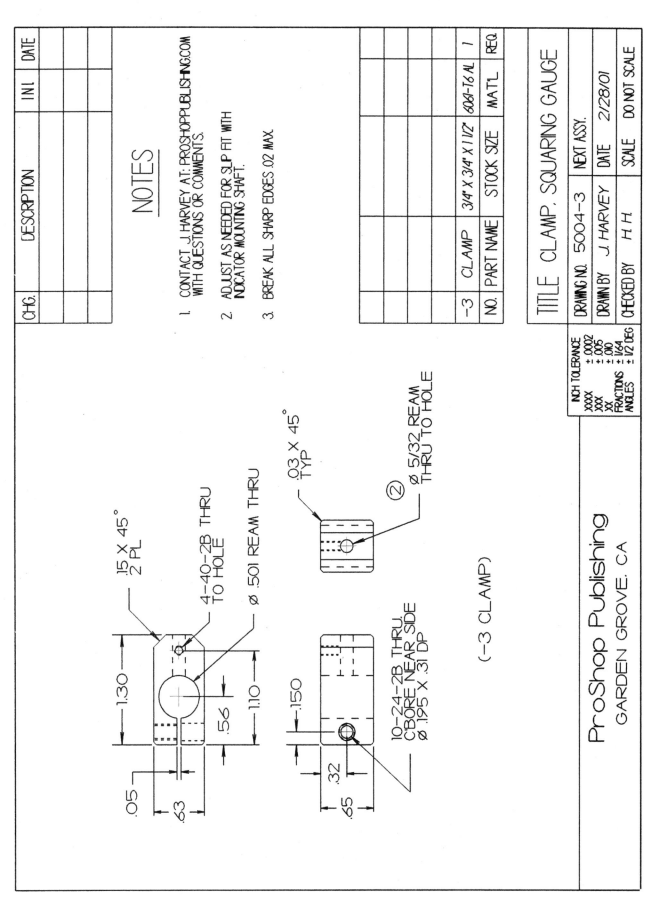

NOTES

1. CONTACT J. HARVEY AT: PROSHOPPUBLISHING.COM WITH QUESTIONS OR COMMENTS.

2. ADJUST AS NEEDED FOR SLIP FIT WITH INDICATOR MOUNTING SHAFT.

3. BREAK ALL SHARP EDGES .02 MAX.

CHG.	DESCRIPTION	INI.	DATE

NO.	PART NAME	STOCK SIZE	MAT'L	REQ.
-3	CLAMP	3/4" X 3/4" X 1 1/2"	6061-T6 AL	1

TITLE CLAMP, SQUARING GAUGE

DRAWING NO. 5004-3	NEXT ASSY.
DRAWN BY J. HARVEY	DATE 2/28/01
CHECKED BY H.H.	SCALE DO NOT SCALE

INCH TOLERANCE
XXXX ± .0002
XXX ± .005
XX ± .010
FRACTIONS ± 1/64
ANGLES ± 1/2 DEG

ProShop Publishing
GARDEN GROVE, CA

.15 X 45°
2 PL

4-40-2B THRU
TO HOLE

Ø .501 REAM THRU

1.30

.56

1.10

.05

.63

.03 X 45°
TYP

Ø 5/32 REAM
THRU TO HOLE

②

.150

10-24-2B THRU,
CBORE NEAR SIDE
Ø .195 X .31 DP

.32

.65

(-3 CLAMP)

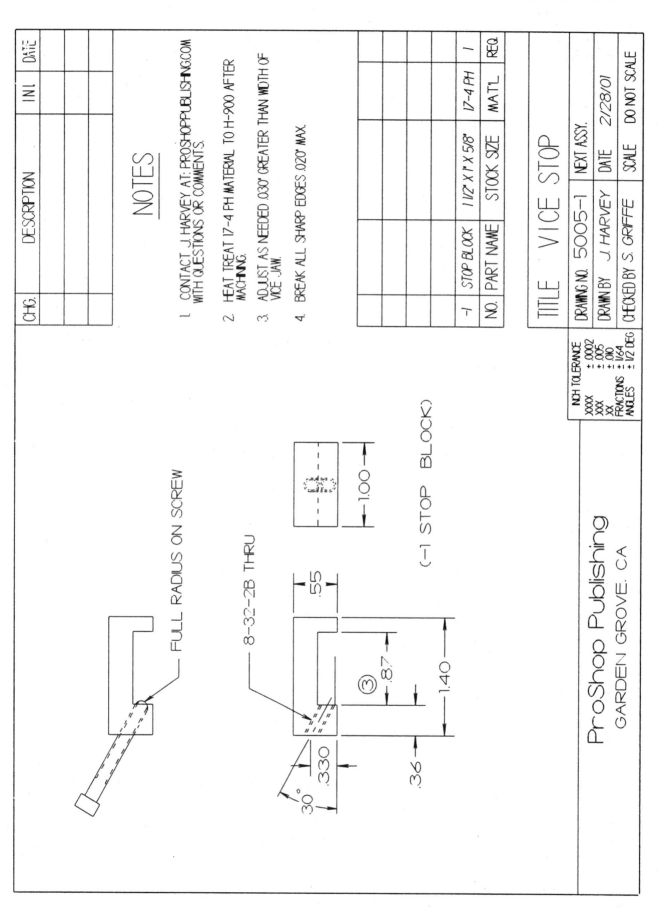

CHG.	DESCRIPTION	INT.	DATE

NOTES

1. CONTACT J. HARVEY AT: PROSHOPPUBLISHING.COM WITH QUESTIONS OR COMMENTS.

2. HEAT TREAT 17-4 PH MATERIAL TO H-900 AFTER MACHINING.

3. ADJUST AS NEEDED .030" GREATER THAN WIDTH OF VICE JAW.

4. BREAK ALL SHARP EDGES .020" MAX.

NO.	PART NAME	STOCK SIZE	MAT'L	REQ.
-1	STOP BLOCK	1 1/2" X 1" X 5/8"	17-4 PH	1

FULL RADIUS ON SCREW

8-32-2B THRU

③

.55

.87

1.40

.36

.330

30°

1.00

(-1 STOP BLOCK)

TITLE	VICE STOP		
DRAWING NO. 5005-1	NEXT ASSY.		
DRAWN BY J. HARVEY	DATE 2/28/01		
CHECKED BY S. GRIFFE	SCALE DO NOT SCALE		

ProShop Publishing
GARDEN GROVE, CA

INCH TOLERANCE	
.XXXX	± .0002
.XXX	± .005
.XX	± .010
FRACTIONS	± 1/64
ANGLES	± 1/2 DEG

Chapter 16 Solutions

1. .0342″ would have to be ground off each wear plate.

2. The ball protrudes .3323″ above the surface of the part.

3. You would have to feed the tap 14.7637 in/min.

4. The min. and max. blade lengths are 143.4″ and 151.4″.

5. The angles of the corners of the pyramid with respect to the base are 35.264°

6. The string would have to be 2π″ or 6.283″ in length. The same is true for any size ball.

7. You would have to add 33 1/3 lbs. of nylon to the existing batch to get the new mixture.

8. See Drawing. There are other similar views that would also be correct involving curved surfaces instead of plane ones.

9. The ladder touches the wall at a height of 9.6771′ above the floor.

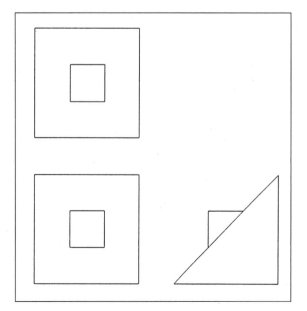

ANSWER 8.

More Trade Secrets

1. *When milling parts requiring multiple setups, begin by making cuts that remove the least amount of material. (See Fig. A3-1)*

   Keep parts as close to block form as long as you can during machining stages so they are easier to hold. Usually you can machine all or most holes first to maintain block form. Then start milling features that remove the least amount of material, which helps maintain block form and part rigidity. Large cuts and angle dressings should usually be done last. This planning technique applies to parts machined conventionally and in CNC machines.

2. *Make a center punch to last a lifetime. (See Fig. A3-2)*

   Start with an old 3/8″ three or four flute tap. Grind the long taper at five degrees per side (ten degrees included angle). Grind the tip angle at thirty degrees per side or your preference. These center punches are compact, durable, and have a great "feel."

FIGURE A3-1   Planning how to run parts is half the battle. A good planning strategy is to begin by making cuts that remove the least amount of material.

3. *Keep your shop stones clean and flat. (See Fig. A3-3)*

   Sand your shop stones over a sheet of 80 or 100 grit sandpaper. Silicon carbide sandpaper works better for abrading and dressing stones than aluminum oxide paper because silicon carbide crystals are harder.

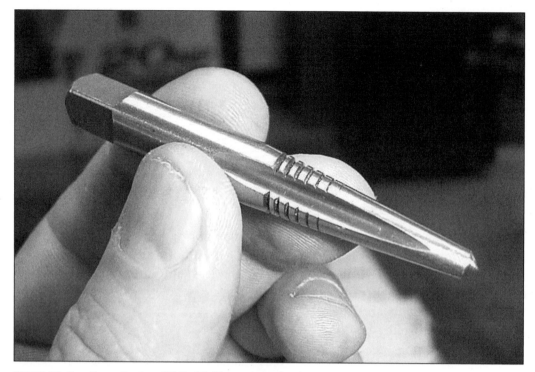

FIGURE A3-2   No matter how "Hi-Tech" things get, you're always going to need a center punch.

FIGURE A3-3   A shop stone is being cleaned and dressed flat in this photo.

Index